The Invisible Killer

The Invisible Killer

The Rising Global Threat of Air Pollution–and How We Can Fight Back

Gary Fuller

MELVILLE HOUSE
BROOKLYN • LONDON

The Invisible Killer

First published in Great Britain in 2018 by Melville House UK

First Melville House printing March 2019

Melville House Publishing
46 John Street
Brooklyn, NY 11201

and

Suite 2000
16/18 Woodford Rd.
London E7 0HA

mhpbooks.com
@melvillehouse

ISBN: 978-1-61219-783-8
ISBN: 978-1-61219-784-5 (eBook)

Design and typesetting by Roland Codd/Beste Miray Doğan

Library of Congress Cataloging-in-Publication Data

Names: Fuller, Gary (Gary W.), author.
Title: The invisible killer : the rising global threat of air pollution : and how we can fight back / Gary Fuller.
Description: First American edition. | Brooklyn, NY : Melville House Publishing, 2019. | First published in Great Britain in 2018. | Includes bibliographical references and index.
Identifiers: LCCN 2019002358 (print) | LCCN 2019004106 (ebook) | ISBN 9781612197845 (reflowable) | ISBN 9781612197838 (hardcover)
Subjects: LCSH: Air--Pollution. | Air--Pollution--Health aspects. | Air--Pollution--Prevention. | Automobiles--Motors--Exhaust gas--Environmental aspects.
Classification: LCC TD883 (ebook) | LCC TD883 .F846 2019 (print) | DDC 363.739/2--dc23
LC record available at https://lccn.loc.gov/2019002358

Printed in the United States of America
1 2 3 4 5 6 7 8 9 10

This book is for the 12,000 people whose lives were cut short by the London smog of 1952. They have no memorial.

Contents

Introduction

Humans can live for three weeks without food and three days without water—but only three minutes without air. Yet we simply take our air for granted. It's always there. It's everywhere. The air pollution that we breathe has changed a great deal over the centuries. It is largely invisible to us but it is having a significant impact on our health and the health of our children.

More than 90 percent of the world's population is exposed to air pollution concentrations that exceed World Health Organization (WHO) guidelines. Globally, four and a half million people died prematurely from particle and ozone pollution in 2015.[1] So why don't we understand air pollution better? And how have we allowed it to build to the crisis we find today?

The face of air pollution has changed. Modern air pollution does not look like the thick black industrial smoke from the past. London's international reputation as the world's most polluted city, beset with pea-souper smog, has been passed to Beijing. We are all familiar with images of Beijing's Bird's Nest Olympic stadium and the Forbidden City shrouded in haze and the city's residents wearing protective masks. Despite this coverage in the news, Beijing does not head the WHO list of the world's most polluted cities. It was 56th in 2016 and dropped to 187th in 2017. Of the worst fifty, the vast majority are in Asia: twenty-four cities are in India, eight are in China, three in Iran and three in Pakistan. Six of the worst fifty are in the Middle East, including four in Saudi Arabia. At the other end of the scale we find small towns in Iceland, Canada, the United States and Scandinavia are some of the cleanest. There are some large cities near the bottom of the list too; including Vancouver and Stockholm, showing that air pollution is not an inevitable part of city life.

As an air pollution scientist at King's College London, my research has focused on the sources of urban air pollution and how these affect people's health. I still lead the London Air Quality Network, the largest urban network in Europe. Over the last twenty-five years I have tracked changes in the air that Londoners breathe, given evidence to the government and worked alongside health researchers and air pollution scientists from around the world. I have measured how London's industrial pollution and problems with gasoline cars have been replaced by diesel car pollution and home wood-burning. Around the world many people look to London's low emission zone as an example of action to control the problem, but if it is so effective then why are Londoners still suffering from poor air? Writing this book has allowed me to explore the real, global problem of air pollution. Expanding beyond my London base I will take you from Paris and Los Angeles to India and New Zealand in a bid to understand modern air pollution. The smog in London and Los Angeles, Scandinavian forest dieback, the Volkswagen scandal and the recent pollution problems across Southeast Asia have all prompted steps to clean our air. We will be exploring the impact that air pollution has on our health; the complex shifting political agenda of air pollution control; the tension between public health and government regulation; and the negative impact of the simple, yet crucial, denial of the problem in the first place. There are huge injustices at the heart of the air pollution problem. By using our air to dispose of their waste, polluters are destroying a shared resource and avoiding the full cost of their actions. They leave all of us who breathe poor air to pay the price through our health and taxes.

* * *

So what do we mean by "air pollution"? Images may instantly spring to mind, such as billowing smoke from car exhaust and chimney stacks. Air pollution comes from many sources, some well-known, such as traffic, industry and coal-burning and some lesser known, including agriculture, wood-burning and volcanoes. Common pollution problems arise from the use of fossil fuels, the pollutants that form in the air around us and natural sources too. At the same time there is huge diversity in the nature of air pollution from place to place depending on the weather, where the air has been before and local controls on the way in which we use our air as a waste disposal route.

You will not need a degree in chemistry or physics to understand this book. It is all about the connection between the everyday pollution sources that we see around us, the air that we breathe and the harm that it does to our health. I will be talking a lot about particle pollution: tiny particles that can be inhaled deep into our lungs. This includes soot from coal-burning and diesel exhaust as well as particles that form in the air from other pollutants. Some pollutants are gases. These include nitrogen dioxide, which is the pollutant at the focus of Europe's diesel exhaust problems, and sulfur dioxide from burning sulfur-rich oil and coal. Ozone will feature in this book too. This gas is better known from the problems of the ozone hole above the north and south poles, but when it forms at ground level it is very damaging to our lungs and affects our food crops too.

Scientists have been investigating the impacts of air pollution since medieval times. Increasingly, we tend to focus on the latest discoveries and findings. The lessons from the past are often forgotten but many of them have huge relevance to the challenges that we face today. I am continually impressed by the insights of scientists who were working with hand-pumped samplers and homemade glassware in their laboratories and calculating their results with slide rules. This book will revisit some of these old investigations and discoveries and tell the stories of the people who made them.

Yet when it comes to the disastrous effects of air pollution on human health, it seems astonishing that insight was sorely lacking for many centuries. This might seem incredible, but it was not until the 1950s that the harm from air pollution was recognized. We are still learning. In 2016, the Royal College of Physicians drew together the latest research to show how the lifelong impacts from air pollution start in the womb and go on to damage children's lungs and shorten adult's lives.

There are many calls for action but fewer examples of positive outcomes in the battle for clean air. Some plans have not worked as well as hoped and many have created new problems. Air pollution is a global challenge that still needs to be tackled alongside climate change and the creation of healthy cities in which to live.

This book starts in medieval London. We will follow the evolution of the way in which we understand the air around us and the warning signs that were ignored. In the 1950s, the deaths of around 12,000 people in the London smog and the eye-stinging Los Angeles air finally brought about concerted actions to control air pollution, which we will explore in the book. We will then focus on the challenges that we face today to ensure that our air is fit to breathe.

Join me on a journey from the smog of the past and present to the hopefully cleaner air of the future.

Part 1

Warning signs: From medieval London to pea-soupers

Chapter 1
Early explorers

You might think air pollution is a modern problem. Or at least one that has developed in the last century. Would it surprise you then to learn that air pollution was being written about as early as the seventeenth century?

It is difficult to imagine life in London many hundreds of years ago. Visits to stately homes and cathedrals show us the buildings of the past but visualizing the daily life of people and the air around them is far harder. In 1661, the diarist and gardener John Evelyn wrote an essay on London's air pollution that he sent to King Charles II and to parliament. His essay was titled *Fumifugium, or, The inconveniencie of the aer and smoak of London dissipated together with some remedies humbly proposed.*[1] The cover letter paints a vivid picture of air pollution at the time (and displays some sycophancy):

> One day, while I was walking in your Majesty's palace, where I sometimes come to enjoy the sight of your magnificent presence, I saw a ghastly billow of smoke coming from one or two tunnels between Northumberland House and Scotland Yard. It was so

> thick that the rooms, galleries and palaces were completely filled with it and people could hardly see each other for the cloud. Indeed, they struggled to even stand up.

London had undergone an energy revolution. The deforestation of the areas around London led to shortages of wood fuel so the city turned to charcoal-burning and then to coal brought in by sea from the northeast of England. This was not the first use of coal as a fuel; a receipt for twelve cartloads of coal was recorded by the monks of Peterborough Abbey in AD 852. But coal had a well-deserved reputation as a dirty fuel; in 1257, Eleanor, wife of Henry III, was forced to leave Nottingham Castle due to the smoke from coal-burning. Previously confined to blacksmiths and lime kilns, coal became the main fuel that powered London in the 1600s. Before this time, when wood was the main home fuel, little attention was paid to the construction of chimneys, but the fumes and soot from coal-burning required more elaborate stacks well above house roofs.[2] The change to the air in the rapidly growing city was plain to see. Evelyn described the center of the kingdom like a scene from Dante's *Inferno*:

> Whilst this smoke belches from their sooty jaws the city of London is more akin to the face of Mount Etna, The Court of Vulcan, the island of Stromboli, or even the very suburbs of hell . . . For, although in other places in England the air is serene and pure, in London the sulphurous clouds are so dense that the sun itself has trouble penetrating it . . . this ruinous smoke that sullies the city's glory, imposing a sooty crust or fur on all the city

> lights, spoiling man's property, tarnishing the plate, gildings and furniture, and corroding even iron bars and the hardest stones because of the caustic elements that accompany the sulphur.

Evelyn was the creator of one of London's finest gardens, Sayes Court, at his home in Deptford, and he could see the firsthand effects of air pollution on the natural environment. He found that London's pollution was

> detrimental to our birds, to the bees and to the flowers, allowing nothing in our gardens to bud, grow or to ripen. Therefore, no amount of work will make anemones, or our other favourite flowers grow in London or the surrounding areas unless they are raised in a greenhouse and carefully nurtured. It is this that means that the few, pitiable fruits that do grow have a bitter and unpleasant taste, and will not reach maturity so that they are like the apples of Sodom that fall to dust the second they are touched.

In a way akin to modern epidemiologists, Evelyn looked at death records to see the impact of London's air on the health of its population.* From 1601, James I required parish clerks to publish weekly lists of births and deaths called the Bills of Mortality. "Searchers," mostly old women, were employed to inspect the corpses in order to establish cause of death. City clerks compiled the information recorded by parishes and sold the bills to Londoners eager to know

* Evelyn was probably referring to the work of John Gaunt, who published his *Observations Made Upon the Bills of Mortality* in 1662, compiling fifty years of weekly bulletins classifying deaths into eighty-one causes. Gaunt's book is discussed at http://www.bmj.com/content/bmj/346/bmj.e8640.full.pdf.

when and where plague was active and therefore the places to be avoided, or when to retreat from the city. City merchant and shopkeeper John Gaunt reduced around fifty years of Bills of Mortality to simple tables of causes of death. Plague years were clear, but it was the constant deaths from chronic diseases that provided Evelyn with evidence of the impact of air pollution:

> Through weakening the people to infections it comes (eventually) to corrode the lungs; this is a problem that cannot be cured and kills scores of people through a long and deep consumption, the proof of which can be found in the city's weekly Bills of Mortality . . . almost half of the people who die in London do so from disorders of the throat or lungs. The inhabitants are never free from coughs or persistent rheumatism, from the spitting up of abscesses and corrupt matter.*

Amazingly, despite this evidence, it seems that the consensus was that smoke was good for London's inhabitants. Evelyn said that he risked "the rejection of a whole faculty, particularly the College of Physicians, who consider it a preservation against infection, rather than the cause of the sad effects that I have described."

The belief that air pollution was a preservative sprang from the miasma theory of disease that prevailed before the discovery of bacteria. Miasma was thought to be an airborne substance produced by rotting and decomposing biological material. Sources of miasma were everywhere. In the countryside, miasma came from marshes

* In the original text these diseases sound even more gruesome: "Phthiscal and pulmonic distempers, coughs and importunate rheumaticms, spitting of impostumated and corrupt matter."

and swamps. In the city, it came from rotting food, horse manure, sewage and even stale breath. A single lungful of miasma was thought to induce zymotic disease, an internal fermentation or rotting which could spread to other people, explaining the clear contagion of some diseases. Fog was thought to be connected with miasma. Both came from marshes and swamps. When plague struck London in the sixteenth and seventeenth centuries people were even urged to light coal fires in the street to drive away the miasma and cleanse the air.[3]

Air was long acknowledged as one of the four elements, along with earth, fire and water, but the concept of the wider atmosphere was not understood. In 1644, the Italian physicist and mathematician Evangelista Torricelli wrote a remarkable letter to his friend Michelangelo Ricci, a fellow mathematician and also a cardinal in Rome, exclaiming: "We live submerged at the bottom of an ocean of the element air." Torricelli had been working on the problem of pumping water from the bottom of deep wells—an impossible task to perform in one action if the well is deeper than about nine meters. Rather than experiment on a full-sized well, he created a small model using mercury instead of water. In his experiment he filled a tube with mercury, placed his finger over the open end, inverted the tube and placed the end in a trough, also filled with mercury. The tube was two cubits long (about 110–120 cm). The mercury did not drain from the tube. Instead it fell part-way down, leaving a vacuum above. This had been done before but Torricelli's insight was to realize that the size of the gap was not a property of the vacuum. The vacuum contained nothing and therefore could do nothing. Instead, "On the surface of the liquid which is in the basin, there gravitates a mass of air 50 miles high."

Torricelli transformed our perception of the air around us. The height of the mercury remaining in the tube was a measure of the pressure from the air above us.[4] This simple barometer could be constructed anywhere in the world and for centuries to come, barometric pressure was measured in inches of mercury.

Just four years after Torricelli's letter, it fell to Blaise Pascal to take the next steps and show that pressure from the atmosphere was not the same everywhere; it decreased with altitude. Frenchman Pascal had been an infant prodigy, particularly in the area of mathematics, but he also worked on the physics of pressure in fluids. You may recall being taught Pascal's law at school. This states that the pressure at any point in a liquid is equally transmitted in all directions. Pascal had the idea of taking a barometer up a mountain. Rather than do this himself he asked his brother-in-law Florin Périer, who lived in Clermont in central France, to conduct the experiment. A group of people met in Périer's garden and filled several mercury barometers. The pressure height of the mercury in the tube was 710 mm. One barometer was left in the garden and watched all day. It did not change. Another was taken to the top of the Puy-de-Dôme, now known throughout the world as a famously challenging climb on the modern Tour de France. At the summit, around 500 fathoms (about 900 m) above Périer's garden, the mercury height was 625 mm; the force exerted by the air above them had dropped by 12 percent. Périer was so impressed by the experiment that he tried it again and again. Perfecting the accuracy of his work, he even managed to measure the reduction in pressure from ground level to the top of the cathedral in Clermont.

At the time the chemical composition of our air was not

understood. Although humanity has used fire for thousands of years, even the essential role of air in burning was never comprehended. If you watch a log fire, the flames seem to leap from inside the wood. They dance in a mesmerizing way and the only apparent function of the air is to fan the flames and to carry away the smoke. This observational perspective was responsible for the theory of phlogiston, a huge fifteenth-century scientific wrong turn.*

Phlogiston was believed to be one of the elements of which matter was composed and was thought to be liberated from a substance when it was burnt. When all the phlogiston was gone, the burning stopped. Flames were therefore not a chemical reaction with oxygen in the air but the excited liberation of phlogiston. Experiments could be performed where mercury could be burnt to ashes, liberating phlogiston, and then the mercury restored to liquid by reheating with charcoal, an obviously phlogiston-rich substance. The tricky fact that the mercury got heavier after combustion, rather than lighter from the liberation of the phlogiston, was oddly overlooked for some time.

It was not until the discovery of oxygen and nitrogen in the 1770s that the chemical exploration of our air was set on course. The air above us determines the pressure but, chemically, is the air different from place to place and from time to time? This was the question that Victorian chemist Robert Angus Smith set out to investigate.

In the preface to his 1872 book *Air and Rain: The beginnings of a chemical climatology*,[5] Smith recounts a conversation with the

* Good defenses of phlogiston theory have been made and neatly summarized at https://thonyc.wordpress.com/2015/10/23/the-phlogiston-theory-wonderfully-wrong-but-fantastically-fruitful/.

physicist and meteorologist John Dalton, who had been experimenting with mixtures of gases. Dalton asserted that "chemical experiment could not distinguish between the air of the city and the air on Helvellyn" (England's third highest mountain). The conversation was a turning point in Smith's science career. Charged with a mission to investigate the chemical composition of air, he conducted a systematic survey of the British Isles. In each location he sealed an air sample inside a glass tube and returned it to his laboratory. He took measurements of the air at the top of Ben Nevis and on the streets of Perth and Glasgow. He also visited Hyde Park in London and pretty much everywhere in between. His research took him as far as Switzerland to measure air near marshes, but the distance seemed to have little influence; as Dalton had suggested, the amount of oxygen changed by less than 0.2 percent between outdoor locations. Smith noted slight differences indoors in hospital wards and cowsheds, but only in the air of mines and chambers where candles were burnt until the air could no longer sustain them did he see differences of whole percentages.

However, he was not convinced that the air was the same everywhere; he theorized that it was the minor differences that mattered, including differences in impurities that comprise less than one millionth of the molecules in our air. With this in mind he continued his investigations. First, he turned to carbonic acid, the acid formed when carbon dioxide is dissolved in water. He found great variation in the amount of carbon dioxide between the boxes in London's Strand Theatre, hospital wards and the second-class carriages of underground trains, on sections of the track that now form part of the Circle Line. He also constructed a sealed lead chamber about the size of a large telephone box at his home and sat inside it for hours, breathing the same air as it changed around him. It took time for him to consume the oxygen; so, impatient for faster results, he persuaded other people to join him, staying until they could barely feel their pulses (his volunteers were rewarded with a good meal afterward, although sometimes they were too ill to eat). Finally, Smith turned his attention to what he termed impurities in the air in towns—what we would now call air pollution. In what must be one of the earliest investigations of climate change emissions, by simple mathematics* he was able to balance the additional carbon dioxide in Liverpool and Manchester's air, when compared to the countryside, with estimates of the tonnage of coal burnt. Importantly, from the air pollution perspective, he found coal smoke also contained many other impurities, including metallic compounds, sulfur, chlorides and acid gases that were polluting Victorian towns. By looking at these Smith was finally able to prove Dalton wrong and show that air did vary from place to place.

In 1859, a House of Commons committee investigation concluded that "the air of large towns had no effects on the lungs when compared with air supplied by nature." Instead, it was argued that it was the differences in living conditions and occupations that caused people in towns to be less healthy and die sooner than people in the countryside. The failure to recognize the health impacts of air pollution will be a recurrent theme in this book. To his credit, Smith set out to investigate if his impurities were linked to health effects. This called for more self-experimentation. Instead of using his chamber, Smith mixed samples of air with diluted amounts of his own blood.

* Dispersion modelers today would term this a box model.

The acid gases found in coal smoke produced a reddening of the blood. With the acid gases removed, the remaining air produced a dull blood sample. Smith thought that redder blood was a good thing and the acid gas impurities, which some considered harmful, might explain the "greater restlessness of the system which is the peculiarity of life in towns." The polluted city air could therefore be important for the vitality of people in our cities and might not be the agent of harm.

This reflected the medical view of the time. Although the miasma theory of disease was being challenged by the discovery of bacteria, it was still the consensus that smoke was a good thing. Just a few years before, in 1848, the surgeon John Atkinson suggested that people with tuberculosis should inhale coal smoke and other chemicals. In his view, creosote, tar, pitch and naphtha could all halt the progress of the disease.

Smith went on to explore rainwater and coined the term acid rain. He saw the way that sulfur-rich rain attacked the stone of buildings, but concluded that it acted as a disinfectant, killing bacteria or even curing disease directly.[6] He also noticed that soils removed the acid from rainwater, making it drinkable again. The damage that acid rain causes to forests and rivers would not be revealed for another hundred years. Smith went on to become the first head of the Alkali Inspectorate, an early regulatory body for industrial pollution. Under his guidance the inspectorate took a gentlemen's approach to enforcing regulations, reasoning that it was better for an industrialist to agree to invest in cleaning his factory's pollution rather than fighting a court case or paying a fine. This view still pervades industrial emissions control today.[7]

Smith was not alone in exploring the air around us. Ten years after the publication of his book, fellow Scot John Aitken set out to investigate soot particles in our air. Aitken was born in Falkirk and had ambitions to be an engineer. Having trained at the University of Glasgow he started work as a marine engineer, but this ambition was thwarted by ill health. Instead he turned to science. He converted his drawing room into a workshop and laboratory, placing a lathe in front of the window and furnishing the room with wooden benches and cabinets around the walls that were filled with thermometers and meteorological instruments. Reflecting his engineering training, his first work on valves was followed by work on the perception of color, but he is best known for his investigation of clouds and fog. To carry this out, he invented a way of making fog and clouds in his drawing room and found, by chance, a way of seeing the tiny particles in our air. Normally particles of soot and other pollutants are too small to see. The wavelength of visible light limits the resolution of even the most powerful microscopes and these particles simply cannot be resolved. However, in his fog chamber he found that each droplet formed round a tiny particle and suddenly they became visible in astonishing numbers: "There may be as many dust particles in a cubic inch of the air of a room at night when the gas is burning as there are inhabitants in Great Britain. In three cubic inches of gas from a Bunsen flame there are as many particles as there are inhabitants in the world."[8]

Aitken's device was simple. He captured air in a chamber and saturated it with water vapor. Suck out a little air to reduce the pressure and tiny droplets form, each one around a particle, enabling it to be seen and counted on a microscope slide.* Aitken's greatest

* Modern particle counters work in much the same way. Air is passed over a butanol wick and the sample

contribution is in our understanding of the way that clouds and fog form,* but, like Smith, he also went out surveying the air. Obviously, the drawing-room sized version of his instrument could not be carried, so instead he created a pocket version, about the size of a cigar case. Aitken counted the number of particles in the air across Scotland and at the observatory on Ben Nevis. Each spring from 1889 to 1891, he went on a European tour to the Alps, Italy, Paris and London, taking with him his new pocket device. In London† and Paris he found between 40,000 and 210,000 particles in each cubic centimeter, very similar numbers to those in London in the early 2000s. Interestingly he also found that air pollution did not stay in towns and cities. Air blowing from the Atlantic was clean, but air blown from cities contained many more particles.[9]

In addition to investigating smoke particles and sulfur, the Victorians also paid a lot of attention to ozone gas in our air. Ozone was first isolated in 1848 by Christian Friedrich Schönbein, a professor of chemistry at the University of Basle. Schönbein made ozone in his laboratory by passing an electric current through water. He quickly recognized it as the smell that remains after a thunderstorm and which today we can find in photocopier rooms.[10] Ozone is a form of oxygen; instead of the usual two atoms found in the oxygen molecules in our air, ozone comprises three atoms. This combination

is then cooled. The butanol condenses on the particles, which grow big enough to be counted with a laser. Butanol is easier to condense than water and is favored for this job, despite its awful smell.

* Oddly, clouds and fog will not form in completely clean air; they require tiny particles to act as condensation nuclei. The high frequency of fog and smog in London and other polluted cities of the nineteenth and twentieth century was in part due to the number of pollution particles in the air.

† Some of his London measurements were made from a window on Victoria Street. A window on the same street was used to measure air pollution during the second half of the twentieth century, including tracking the 1991 smog.

is unstable, making the ozone molecule a powerful oxidant, prone to react with many substances, including the insides of our lungs, in an effort to ditch the extra oxygen atom.

Ozone was not always recognized as harmful. In Victorian times, taking the air and breathing the ozone was thought to be part of the health-giving properties of a visit to the British seaside. Even today you can enjoy an ice cream while admiring the sea view as you walk along Ozone Terrace in Lyme Regis, Dorset. However, the linkage of ozone and the seaside probably results from confusion about smells that was seized upon by coastal towns as a marketing ploy to encourage visitors. Schönbein placed a lot of emphasis on detecting ozone by smell. Yes, the seaside smells like ozone, but what we smell is generally a gas produced by bacteria and microbes that live on seaweed rather than ozone itself.* In reality, ground-level ozone is very harmful and has no place alongside the joys of water parks and sandcastles on the beach. Even in the mid-1850s scientists had discovered that breathing large quantities of ozone caused chest pains. They knew that rabbits and mice died quickly when they breathed the gas,[11] but the view that ozone was good for us prevailed, again due to the miasma theory of disease.

One experiment showed how these clearly contradictory views on the health impacts of ozone could be unified by the miasma theory. In 1866, the physician and public health expert Benjamin Ward Richardson was investigating miasma. His source was an eight-year-old flask of rotting oxblood that produced what he

* More recently, the smell of the seaside has been attributed to dimethyl sulfides from seaweed and salt marshes: http://www.uea.ac.uk/about/media-room/press-release-archive/-/asset_publisher/a2jEGMiFHPhv/content/cloning-the-smell-of-the-seaside.

described as a nauseating smell, the nature of which I will leave to your imagination. When mixed with highly reactive ozone, the odor went away. Today, highly reactive ozone is used in some kitchen extractors to remove cooking smells, but Richardson interpreted this as a destruction of miasma and became convinced that ozone would improve the air of our cities. He went as far as suggesting that an ozone company should be set up to pipe the gas into butchers' and greengrocers' shops to preserve meat and vegetables and to provide seaside air to every home.[12]

Further evidence of the curative powers of ozone came from an outbreak of relapsing fever in the St. Giles district of London. Spreading rapidly in damp, overcrowded boarding houses, relapsing fever was thought to be a zymotic disease spread by miasma. Dr. George Moss, medical officer for the area, suggested that outbreaks of relapsing fever increased when the ozone level in the city fell due to coal smoke and smog. We now know that relapsing fever is spread by lice and tick bites and has no connection with air quality, but the relapsing fever evidence fed into the prevailing narrative that the risk from air pollution came from the offensive odors of rot and decay.

Ozone was one of the first pollutants to be measured routinely in our air. Starting in 1876, daily measurements of ozone were made for thirty-four years at the Observatoire de Montsouris, then in the outskirts of Paris. Each day a sampler was placed on the balcony of the house and air was sucked through a liquid reagent. The results gathered dust in the statistical bulletin of the city of Paris for nearly one hundred years before their rediscovery by German scientists Andreas Voltz and Dieter Kley. Making sense of the results was not easy. It required Voltz and Kley to build their own version of the sampler from the original drawings and painstakingly recreate the laboratory experiments. What they found astonished them. Comparing the Paris measurements from the late 1800s with those obtained in the twenty-first century revealed change on a massive scale. The average concentration of ozone that we breathe today around the world is more than double the ozone our ancestors breathed just over one hundred years ago.[13]

Smog and fog became a defining part of London's character in the nineteenth century. They feature in the Sherlock Holmes books, those by Charles Dickens and many other authors of the time. They were very different to normal fog, not least because of their color, which has been variously described as yellow, brown and orange, giving rise to the term "pea-souper." The artist Claude Monet came to London between 1899 and 1903 to paint the smog-shrouded city. Many versions of his sunsets over the Houses of Parliament are on display in galleries around the world, and his paintings of Waterloo and Charing Cross Bridges reveal skies filled with black soot and gray and yellow hues.

The smog was certainly disruptive. During the thick smog of 1873, fifteen people were reported to have drowned at Northside Docks, two men walked into the river at Wapping and two workmen died falling into the Regent's Canal. Numerous tales tell of people who got lost having disembarked from their carriages to lead the horses, but were then unable to find the horses, the cabs or any landmarks around them. Others got lost right after emerging from their own front doors. Smog was not confined to London. Smith reported in the 1880s, for example, that "the smoke of Manchester had little exit from the town . . . The eyes began to smart and in walking on pavements cartiers were met leading their horses into shops in the daytime—we can scarcely say in the daylight."[14]. Needless to say the thick fogs were a great opportunity for pickpockets and thieves too.

So, as the Victorian era drew to a close, our knowledge of air pollution was gaining ground rapidly. For the first time, measurements had been made of so-called impurities in the air and how these changed from place to place, but less was known about how they changed from time to time. The European tours of Aitken and Smith were mainly taken in the summer, but what was air pollution like in these places in the winter? Were some cities worse than others? These questions only began to be addressed in the following century. One important lesson from these early explorers of our air was that pollution did not stay confined to towns and cities; a lesson that was forgotten in the years to come, as we shall see in Chapter 6. Despite the inconvenience of smog and the irritation of the eyes and throat, the prevalence of the miasma theory of disease and the perceived disinfectant properties of some of the nastier pollutant gases meant that air pollution was perceived as irritating but harmless, or even beneficial to health. The true nature of smog as an invisible killer would not be recognized for another fifty years.

Chapter 2

Warning signs ignored

In many respects air pollution in the UK during the first half of the twentieth century was a continuation of the Victorian era; the growth of air pollution in our cities still plotted much the same course, rather like a ship heading toward an iceberg. Into this collision course stepped John Switzer Owens, who, more than any other person, would define the transformation of air pollution science from the haphazard investigations of Victorian gentlemen into a systematic national surveillance program. Owens did not sit quietly on his findings. He ensured that evidence on the state of our air was heard as widely as possible through his work with anti–coal smoke campaigns, books, talks to scientific societies and reports to government. The science journal *Nature* described him as "a most useful and public-spirited man of science" who was "the moving spirit in the investigation of atmospheric pollution . . . Owens' inventive skill, contentious compilation of records and personal enthusiasm resulted in Great Britain being far ahead of any other country in the knowledge of the pollution of its atmosphere."[1]

Owens was born in Enniscorthy, Ireland, in 1877. He was uniquely skilled, being trained as both a doctor and an engineer. He started his career with a medical degree at Trinity College Dublin, but gave up medicine after five years to become an engineer and work on coastal defenses and mining equipment. In 1912, an international exhibition was held in London under the auspices of the Coal Smoke Abatement Society, of which Owens was a member. Set up in 1898 and rooted firmly in social reform and Victorian philanthropy, the society was well connected with the influencers of the time. National Trust founder Octavia Hill had been motivated by the contrasts between the clean air seen during her trips to Nuremberg, Germany, and the pollution prevalent at home in Britain. Playwright Sir George Bernard Shaw was also connected to the society, and addressing a meeting on one occasion he explained that the secret of health and cleanliness was a clear atmosphere and clean clothes. With these "you will live as you do in the country, where you never wash at all, except as a sort of social ceremony to prove that you are well brought up." Having undergone various name changes, including the National Society for Clean Air, it continues today as Environmental Protection UK, the world's longest running environmental campaign group.[2]

A public exhibition and conference to promote awareness of air pollution in 1912 led the UK Meteorological Office and various municipal authorities to form the Committee for the Investigation of Atmospheric Pollution. This was a voluntary organization, supported by *The Lancet*, and Owens became its first secretary. Initially working for free, Owens was employed part-time as superintendent in charge of measurements when the committee's work was reassigned to the Met Office in 1917, and ten years later it became his full-time job when the committee's work was taken up by the government.

Standardizing measurements across the country was an early task. Bizarrely, the first standard method for measuring airborne particle pollution had little to do with the actual pollution in the air.* Instead the pollution that fell to the ground was collected and weighed.[3] The original idea came from scientists seeing the soot that fell on clean snow in Manchester in 1902, reinforced by later experiments that collected dirt and dust in boxes distributed around Glasgow in the winter of 1906. Owens' collector was perfected in trials at four locations in London. Following the Victorian tradition of home experimentation, one of the trial locations was at his own house in Cheam, now a suburb of south London.[4] The fictional suburb of East Cheam will be familiar to fans of the comedian Tony Hancock; Owens lived in North Cheam, in the idyllic-sounding Daphne Cottage on Wordsworth Drive.

A national network of so-called deposit gauges was set up in the UK in 1912, and by 1936 measurements were being taken in over 150 places. These were very basic: a funnel collected dust that was then washed into a collection bottle beneath. The device therefore measured anything that fell to the ground, including soot and dust washed out by rain. In the days before plastic, the acidity of the polluted air and rain caused problems. Early gauges corroded badly and had to be enameled, leading to distortions in shape. Small

* You would think that measuring the amount of particle pollution in the air would be easy. It should be straightforward to weigh a sample on filter paper, but in the early twentieth century sufficiently accurate scales were not generally available. Other tricky problems included the tendency for filters and the collected particles to take up and lose water vapor, affecting their mass. Filters would also gain static electric charge and simply float above the scales, stubbornly refusing to be weighed. A network of measurements in lots of towns and cities was never going to work using this method.

wire fences were also fitted around the edge of each gauge to prevent birds perching on the instrument and then defecating into it. Other difficulties were found: children threw stones into the collectors and drunks urinated in them on their way home from the pub. Positioning the gauges out of harm's way became very important.

A huge amount of dust collected in Owens' gauges. Hundreds of tons of soot and dust fell on every square mile of cities and towns. London was in fact not the most heavily polluted location. Notable places at the top of the pollution league included the glass manufacturing town of St. Helens in northwest England, where an astonishing 685 tons of dust and soot fell from the sky on each square mile in 1917. This is around half a pound on each square yard. Householders would have been dealing with piles of soot on their doorsteps, in addition to downdraft from their own fires, making it a huge task to maintain a clean home. This was not just a problem in industrial towns. The results for Liverpool, some fifteen miles from St. Helens, were much the same.

Malvern, a small spa town in Worcestershire, was used as the benchmark clean place against which to compare all other British towns and cities. Malvern experienced between one-fifth and one-tenth of the soot and dust that fell from the skies on the average town.

Owens' results were not always well received. He was criticized by his peers for expressing his results in ways the public might understand. Having addressed the Society of Public Analysts in 1925, a Mr. R. C. Frederick accused Owens of sensationalizing the problem: "The expression of the results in tons per square mile, while useful for propaganda purposes, gave, from the scientific point of view, an altogether exaggerated impression of the condition of affairs."[5] Despite this, deposit gauges developed from Owens' work are still used around the world today, mainly around quarries and mines.One major weakness of Owens' approach was that only the largest particles of soot and dust would land in the gauges, having fallen to the ground near big chimneys. By collecting only the pollution that fell to the ground, Owens was focusing on very local pollution sources. This built into a general understanding that air pollution was solely a problem for towns and cities and not for the areas around them. Owens carefully tracked the measurements, and by 1936 he was able to report the results of twenty-five years of policies to curb air pollution. They were not very impressive. Although air quality had improved in London, Glasgow and some big cities, the conditions across the industrialized north of England, including Liverpool, Stoke on Trent, St. Helens and Leeds had become worse.[6] The framework for air pollution control, with a focus on controlling air pollution from industry, was not working.

Owens' second invention reinforced the view that towns and cities were the main sites of pollution. This device used a water siphon to suck air through a white filter paper. Early versions used water from large, sealed glass jars. As water drained from the jar it drew air through the filter paper, turning it gray or even black depending on the soot in the air. Later versions automatically refilled and then flushed every hour using a timer and an arrangement not unlike a toilet cistern. In this way, air pollution measurements could be made each hour. Staff at each measurement station would look up the shade of gray on a chart and read off the corresponding amount of particles in the air. Finally, electric pumps were used instead of water siphons. Owens' invention became known as "British black smoke" and would play a major role in the study of pollution for years to come.

Interestingly, although legislation and pollution control had focused on industry, Owens' measurements often showed that something else was the problem. In many towns and cities, it was home fires that dominated air pollution. The state of the UK's air was the subject of a government inquiry starting in 1914. Government had been shamed into a response by Lord Newton, who proposed a new law on smoke control. Rather than supporting or blocking the proposal, the government agreed to let Newton head an inquiry, thereby kicking the problem into the long grass. The First World War intervened and the committee did not report until 1921. In a precursor to air pollution debates today, Newton's committee said that "the chief factor in the failure to deal with the smoke evil has been the inaction of the Central Authority." It called for better controls on industry and for controls on the exhaust and smoke from trains, cars and trucks, with stronger fines and new standards defining what controls were "practicable." Apart, however, from calls for research and for new houses to be designed with smokeless heating, the committee proposed no actions on home fires. Owens criticized this inaction in light of the clear experience of everyday life in London and other cities:

> We are all personally conscious of the exceptional impurity of the atmosphere by the smarting of our eyes and the annoyance to our breathing, by the collection of dirt in our nostrils, on our clothing, on curtains and furniture, and by the more permanent record in the deterioration of buildings and metals. On special

> days, such as Tuesday, November 19th, 1922, London paid the penalty by living in darkness the whole day.[7]

As we will see, government has always been reluctant to act on what we do in our own homes, even if it impacts the health of other people. Lord Newton's report was greeted with complete indifference. *The Times* was pleased with the report as "a sane and convincing presentation of a complex problem the more weighty because it suggests no heroic measure." It was business as usual. The industrialist Sir Hugh Beaver, who would head the next government inquiry in 1954, noted that although Newton's report ran to over 860,000 words, nothing happened until the Public Health Act fifteen years later in 1936; even that was "full of loopholes and reservations" and "failed to achieve its purpose."[8]

Owens invented a third instrument to measure air pollution in 1921 and discovered something that sadly remained largely overlooked for the next fifty years.[9]

Like all good inventors of the Victorian and Edwardian period, Owens took his experiments on his summer holidays. He visited Holme on the Norfolk coast of England and amused himself measuring air pollution on the beach. Holme was far away from large towns, cities and industry, and the summer sun meant no home heating was being used. What he found was a real surprise. It seemed clear that if you relished clean air, you needed to be in the countryside. On this basis, Owens would have expected his coastal holiday air to be very clean.

Owens' new instrument did something different than his other two inventions. Instead of measuring the big soot particles that fell to the ground, or the black particles only, it measured all the particles in the air. Air was sucked into the device using a piston resembling a bicycle pump. It passed through a damp chamber and was then accelerated close to the speed of sound before being smashed onto a microscope slide. A few swift pumps, generally between two and twenty, and the sample was complete. Owens would pop a glass cover on the slide and take it home to study under his microscope.

To his surprise, the bluish haze on his otherwise clean Norfolk beach was full of particles. There were hundreds in each cubic centimeter. But where did they come from?

To solve this puzzle, Owens started with weather data. Being a practical man, he took wind measurements by timing the drift of thistledown across fifty yards of sand.* When he returned home, he combined his own measurements with those from nearby coastguard stations and was able to track the wind backward. On some hazy days in Holme, winds came from the industrial areas of the Midlands and Yorkshire, but on the most polluted days the winds blew from the continent—and most especially from the industrialized areas of Germany, around 300 miles away.

Armed with this new knowledge that air pollution was not confined to cities, he went out to investigate further. He found pollution in Devon that had traveled 185 miles from London, and pollution on the southern coast of England that came from the Midlands. He even discovered a daytime haze "as thick as fog" at St. David's Head in Pembrokeshire, west Wales, that came from London, 250 miles away.[10]

Owens' instrument also allowed him to measure the size of

* It is unclear if Owens was assisted by family members. He and his wife did not have children of their own but chasing thistledown could have been great entertainment for kids.

particles in the air.* He calculated that, thanks to their tiny size, they could be in the air for five to ten days, enabling them to travel long distances. The type of pollution that he measured on his holiday was not the same type that was falling by the ton onto nearby towns; it was something different. It was hard to believe that all the particles were coming from industry. However, these haze events were downwind of industry and cities, so coal-burning had to be the cause in some way. The coal strike in the summer of 1921 supported this theory. With no coal being burnt, people could see that the world around them was miraculously transformed, not just by the clearing smoke but because distant hills and towns were visible in a way that people had never seen before.[11]

Owens returned from his Norfolk holiday to his home in Cheam, but he continued to collect samples each day. Something very odd began to happen the following March.[12] The number of particles in the air began to increase, but again it was not the peak home heating season and the particles were not sooty black. By the end of the month over half the particles polluting London's air were transparent and not soot.

Owens set off looking for the source. First, he thought that it was close to his home, so he took his instrument to his office in central London, nine miles away, to triangulate the source. Oddly, the same particles were detected there too. Next, he traveled to St. Albans, north of London, around forty miles from home, and again he found the same particles. The pollution was clearly covering a very large area.

* These were mainly between 0.5 and 1.5 microns in diameter. A micron is one millionth of a meter or one thousandth of a millimeter, about 1/50th the diameter of a human hair.

Looking more closely, the particles appeared to have dots at their centers surrounded by a transparent coating. Many ideas were put forward by the scientific community at the time. These included plausible suggestions such as pollen, spores and bacteria, drops of oil or furnace ash. Other ideas included debris from volcanoes, cosmic or solar dust from space or particles produced from unspecified radioactive interactions of carbon dioxide and water. In summary, the source remained a mystery.

Another addition to the puzzle came from the Kew Observatory, a grand white house in suburban west London not far from the world-famous botanical gardens. The house had been an observatory since 1767 and was leased to the British Association for the Advancement of Science as a laboratory for the physical sciences.* In the late 1920s, Kew scientists noticed that it was not just the streetcars that were messing up their measurements. Their careful measurements of the electrical properties† of the air changed through the day in a cycle that almost, but not quite, matched the changes in soot particles that they measured using one of Owens' instruments.[13] However, these electrical measurements told them something more; that "other" particles were still there on days when no black particles were found. This was much like Owens' seaside and spring particles. Even more strangely, there was evidence in the data that some of the particles might have disappeared in the middle of the day. This was certainly not consistent with the idea that urban smog consisted

* The gradual expansion of London and its streetcar networks nearby led to the observatory's eventual closure in 1924. However, part of its legacy, the UK's National Physical Laboratory, still thrives as a world-leading measurement institute just a few kilometers away in nearby Teddington. See http://www.geomag.bgs.ac.uk/operations/kew.html.

† They measured the potential gradient. You can find out more on this by reading Richard Feynman's lecture at http://www.feynmanlectures.caltech.edu/II_09.html.

simply of soot from coal-burning. Clearly the particles in the air were much more complex.

Owens devoted a large part of his 1925 book to his third instrument.[14] It was featured in his academic paper "Suspended impurity in the air" with detailed technical drawings so that anyone could make their own. He lent his instrument to scientists around the world. Measurements were made in Portugal, Greece, Australia and North America and were also taken from aircraft and balloons.* However, with each use, the instrument seemed to pose more questions than it answered. Perhaps for this reason, it never caught on. Instead, Owens' deposit gauge and the British black smoke method that he pioneered became the standard methods for measuring air pollution across the UK and around the world. The black smoke method was developed further and would be codified as an international standard in Paris in 1964. I cannot think of another instrument that was so important in the battle against air pollution.

Despite many advances in medicine, knowledge of the harm caused to health by air pollution was only creeping slowly forward. Addressing a meeting of the Smoke Abatement Society in Manchester, Dr. J. S. Taylor, the city's assistant medical officer, summed up the state of knowledge in 1929.[15] Like others before him, he drew upon differences between death rates and infant mortality in town and country. Data was presented that showed that "black fogs" in the winter of 1923 to 1924 were accompanied by increased deaths from respiratory disease. However, winters were not just foggy, they were also cold, and it was the low temperature that was

* Not all of these sampling trips ended well. Two scientists sadly died taking samples when their balloon was struck by lightning.

thought to have caused the deaths. Taylor explained that the main linkage between ill health and smog was not through breathing it in but through the darkness it caused in our cities.

Lack of daylight was a very real phenomenon and it is hard to imagine now what it must have been like to be under permanently dark skies. Victorian meteorologists often reported what they called high fog, which blocked the sun. Between 1881 and 1885, central London had only 17 percent of the winter sunshine measured in the countryside, although this improved to 45 percent by 1916 to 1920.[16] A further curious phenomenon was sudden daytime darkness, when the skies would rapidly blacken and the city would become as dark as midnight. Daytime darkness was recorded in the winters of 1912 and 1924,[17] creating huge problems for early power plants that struggled to cope with the sudden demand for electricity caused by the use of lighting.*

This darkness was thought to affect health in two ways. The first was to cause a lack of vitamin D, resulting in numerous cases of rickets in both children and adults. The second impact was more direct. Helio (or sunlight) therapy was being hailed as a successful cure for tuberculosis[18] at the time (in a pre-antibiotic age there were few options) and therefore the lack of sunlight in smoke-filled cities was thought to explain the respiratory disease.† So yes, it was important to clear the skies, but there was less worry about breathing air pollution.

* A high fog in 1955 caused the skies to go completely black at one o'clock in the afternoon. Peter Brimblecombe, in his excellent book *The Big Smoke*, Suggests that smoke from the city was lifted high by the Chiltern Hills (to the north of London) and flowed back over the city. In 2017, a combination of Saharan dust and smoke from Portuguese forest fires caused daytime darkness over much of southern England. See https://www.theguardian.com/uk-news/2017/nov/12/pollutionwatch-sepia-skies-point-to-smoke-and-smog-in-our-atmosphere.

† Although I present this to illustrate that harm from breathing smoke was not on the radar in the 1920s, emerging research today is linking insufficient vitamin D to respiratory illness.

In addition to the diligent measurements collated by Owens, further warning signs came from a series of tragedies over the next two decades that brought people face to face with the deaths caused by air pollution.

The first lethal event occurred in 1930, in the narrow, industrialized Meuse Valley, between Huy and Liège in Belgium.[19] A dense winter fog trapped smoke in the valley for five days. Several hundred people suffered from breathing problems and sixty died in quick succession. Their demise was horrific; painful chests, coughing and breathlessness followed in some cases by foaming at the mouth and vomiting before total heart failure. Many cattle had to be slaughtered and others were only saved by herding them up the hillsides, above the fog.

The investigators speculated about what had caused so many people to die. Poison gas weapons left over from the war were suggested as a possible cause, but this was rapidly dismissed. Investigators found that there had been two earlier such incidents. Cattle had died in a previous fog in 1911 in a side valley. In another, hydrofluoric acid gas had leaked from a fertilizer plant in the valley, causing damage to windows and light bulbs, but a repeat of this would not have been enough to poison the whole valley. Finally, the finger was pointed at the industry in the valley. They concluded that the sulfur and soot from the huge amount of coal burnt there was the most likely cause. The investigation contained an ominous warning that a similar incident inflicted on a large city population could cause many more deaths. The prediction was that the same scenario in London could kill as many as 3,200 people.[20]

The second incident occurred eighteen years later in the US town of Donora, Pennsylvania. Located in a valley thirty miles south of

Pittsburgh, the town was dominated by a steel blast furnace and zinc works. A two-day smog in 1948 led to the deaths of eighteen in a community of 14,000 people, with many more reporting breathing difficulties.[21]

A third incident happened just two years later in Poza Rica, in the center of Mexico's main oil production region. A new plant to remove smelly sulfur compounds had been put into operation, despite only being half complete. A flaring system, set up as a temporary measure to burn off poisonous hydrogen sulfide, failed after two days and the gas was emitted into the windless city air. Twenty-two people died and 320 were hospitalized. The investigation concluded with views that seem rather understated. It recommended that alarms be installed around the plant to prevent a recurrence, and that "industrial health and safety provisions might well be improved."[22]

The relevance of these incidents in helping to understand urban air pollution was, however, dismissed. All the locations were heavily industrialized. Although the nature of the poison gas in the Poza Rica incident was identified by the plant operators, the investigators in the Meuse Valley and Donora had no pollution measurements to tell them what the harmful agent was. The warning signs were ignored.

Part 2

The battle begins: Twentieth-century damage

Chapter 3

The great smog

Today, images of smog from cities across Asia, including Beijing and Delhi, are becoming ever more commonplace, but these smogs look like haze and are caused by particle pollution. London's infamous smog was different. It was a combination of water droplets with the soot and sulfur from coal that powered London's industry and kept homes warm. It is difficult to imagine being outside in fog so thick that you could not see your feet. It is equally difficult to know how it felt to be at home, struggling to keep warm around the fire with fog pressed against the windows as thick as night, even during the day. Fog seeped indoors too. John Switzer Owens said that smogs caused a haze that he could see looking across his living room, but the smog made itself obvious in any large indoor space, such as a theater or a cinema.

On Friday, December 5, 1952, my dad left work in south London, stepped out into the darkness and realized this was not going to be an ordinary journey home. The fog was exceptionally thick. It was as if the world around him had vanished. Being seventeen and youthfully undaunted by the risks, he set off on his one-mile bike ride home. By carefully following the line of the curb, it was possible to count the road junctions to navigate, but halfway home he found the road blocked by a delivery truck from the local laundry. The driver was completely lost, unable to see anything. My dad did his best to help, walking in front of the truck to guide the way. He could just about find the curb by sight and touch, while the truck driver could see the bike's light and hear my dad through the open cab window. Slowly they made their way along. By the time they found the laundry, less than a half mile away, the driver was two hours overdue.

That Friday was clearly not going to be a night at the pictures with my mum. The dreaded notice scrawled on the blackboard would inevitably declare "fog in the cinema" so that no one would be able to see clearly. As a mischievous young child my dad had often been responsible for allowing fog to gather inside the cinema; sneaking down to open the fire doors, he would watch with amazement as the clouds rolled in, before he was chased away by the ushers and usherettes. That Friday night the smog had a serious impact on London theaters. The opera *La traviata* had to be abandoned at Sadler's Wells after the first act because the theater was so full of smog, and at the Royal Festival Hall, those with balcony seats could not see the stage. Weekend football matches had to be abandoned.[1] The *Daily Telegraph* reported the cancellation of a live broadcast by the BBC Symphony Orchestra and the cancellation of a radio concert after the lead pianist got lost on the way to the venue.

The first news articles on the health impact of the fog reflected the British love for animals. On Saturday, December 6, the *Daily Telegraph* reported that a duck had flown into a Mr. John Maclean in the thick fog in Fulham, west London. Both were slightly injured, and the duck was taken to the vet at the RSPCA. However, it is the news reports of events at the Smithfield farming and agricultural show that gave the first warnings of something more serious. Animals began arriving on the Friday to be settled in for Monday's show. In scenes reminiscent of the Meuse Valley smog twenty-one years earlier, the cows started to develop breathing problems as the smog closed in, panting with their tongues out. The official report devotes an entire appendix to the effects of smog on the animals at the Smithfield show. One hundred cattle needed treatment, sixty needed major veterinary attention, one died and twelve more had to be slaughtered on humane grounds. Oddly, neither sheep nor pigs suffered any ill effects from the smog.[2]

As the weekend slowly progressed, the UK's greatest peacetime disaster unfolded. The 1952 smog caused death rates in London to increase by 2.6 times.[3*] A doubling or tripling of the number of deaths that we hear about in our own communities for a week would be unnoticeable. We might not know anyone affected. If we did it would have taken time for the news to arrive by word of mouth, when people were trapped indoors by the smog in the era before the internet and home telephones.

Hospitals were the first to realize that something serious was happening. Slowly they became overwhelmed as the ill were brought in through the dense fog. Soon, almost 500 people per day were being moved around the capital in the search for hospital beds, and this continued into the following week. The *Telegraph* reported the

* Several papers and reports were produced in the aftermath of the 1952 smog. These covered aspects including the air pollution, meteorology and health effects. The Ministry of Health's 1954 report *Mortality and Morbidity During the London Fog of December 1952* draws these evidence strands together and is by far the best reference. Sadly, paper copies are rare and it is not yet available online from the National Archives. Thanks are due here to my retired colleague Steve Hedley who loaned me his copy.

pressures on the ambulance service. It took five or six times as long as usual to travel to hospitals and women—including the wife of Selwyn Jones, a well-known football player at the time*—gave birth to babies in fog-bound ambulances.

When the coroners' offices opened on Monday they were overwhelmed by the weekend's deaths. The 1954 Ministry of Health report comments, "In view of the large numbers of bodies on Monday 8th December, the pathologists had little time for detailed examinations."[4] News reached the ministry and local medical officers were ordered to search for signs of an outbreak of infectious disease. Locations and causes of death were tabulated and mapped and staff made door-to-door enquiries. They found evidence of breathing problems, heart attacks and strokes; not unusual for wintertime London in itself, but the sheer rise in the number of cases was cause for concern. Importantly, there was rarely more than one death per household. Infection was plainly not the cause.

It stayed foggy all weekend, and the fog was still thick when my dad returned to work on Monday morning. He was an apprentice French polisher for an undertaker in south London. Mondays were always the busiest day of the week, as this was when coroners released the bodies of those who had died over the weekend. A busy Monday was termed "a double-digit day." But Monday, December 8, was like no other. With bodies filling the hospital mortuary, the normal method of collecting corpses, one by one in a discreet "private ambulance," simply was not viable. Instead a wood truck had to be used, on one occasion returning full with eighteen bodies.

Some breaks in the fog appeared on that Monday, and it finally

* He played outside right for Leyton Orient.

dispersed on the following day. The fog had covered around a thousand square miles.[5] The deaths were spread thinly but the body count was vast. The Ministry of Health report observed:

> Only those immediately concerned with death . . . were able to realise and then only on a local basis, how great the mortality had been. It must be in truth a supreme example of the way in which a metropolis of eight and a quarter million people can experience a disaster of this size without being conscious of its occurrence. Not until the death certificates were assembled and analysed did the extent of the excess mortality become apparent.

The smog was officially estimated to have killed close to four thousand people, mainly the very young and those over forty-five.

Figure 1 Daily deaths for Greater London in the 1952 smog and pollution concentrations

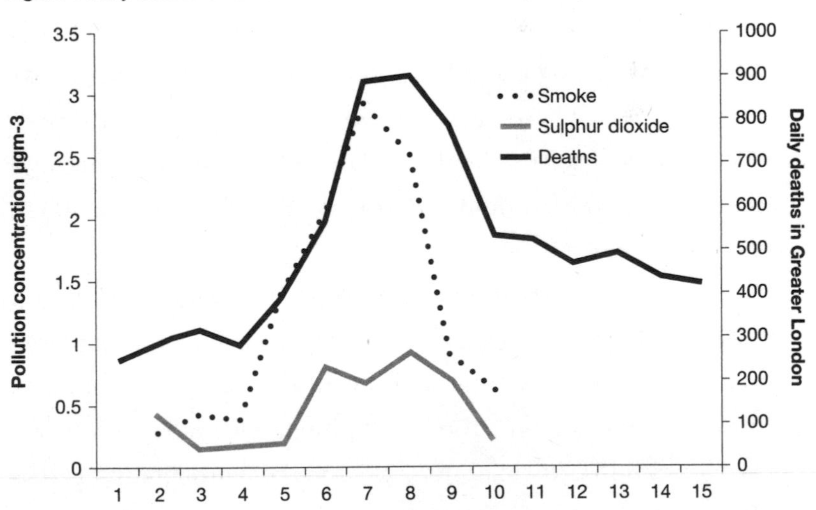

By 1952 a network of pollution monitoring stations had been set up across the capital, their operation based on Owens' black smoke method. The samplers also measured sulfur dioxide. Figure 1 shows air pollution measurements during the smog, plotted against each day's deaths.[6] It is clear that the death rate directly spiked as air pollution increased—but the worst was yet to come. The death rate remained high after the smog cleared, and this persisted through to the following March. Although experts at the time discussed the possibility that the fog had left people weak and vulnerable, having broken news of a massive death toll from air pollution, the Ministry of Health committee simply could not believe that the sudden effect of the smog would last for so long. This was hotly debated. Two months after the smog, a fatality figure of 2,851 was announced by the Minister of Health, only to be revised to 6,000 just two weeks later.[7]

Eventually the government's chief medical officer decided that only deaths before December 20 would be counted. The rest had to be due to something else, most likely influenza. It simply was not credible that the smog had killed more than four thousand people. Not everyone agreed. The government's chief air pollution expert, E. T. Wilkins, focused on the elevated death rates that persisted and pointed out that these would have added around an extra 8,000 to the official figure.[8]

To mark the fiftieth anniversary of the 1952 smog, the health records were looked at again, and the influenza idea was explored in detail. Even with a worst-case influenza epidemic the numbers did not stack up. In 2002, the best estimate of the death toll was revised upward to a tragic 12,000,[9] close to the figure first suggested by Wilkins.

The week after the smog hit, an extra 15,000 people were registered as being too ill to work and therefore eligible for government sickness payments.[10] This is likely to be a substantial underestimate of the number of people made ill by the smog. In the earlier Meuse Valley, Donora and Poza Rica incidents, the number of people left ill was much greater than those who were killed. People who go to work each day are normally the healthiest in the population and would be expected to be the most resilient. There are no statistics on illness in the very young and old, but the numbers left ill would have been far greater than the 15,000 who became too ill to work.

But why did so many people die in 1952? And why had this not happened on such a scale before? Several possible causes have been suggested. Fingers were pointed at the Thameside power stations at Bankside and Battersea. Attempts at mapping the smog found high concentrations near these power plants, but observations from aircraft had noted that the chimneys were often above the smog layer. The quantity of coal being burnt in city homes was blamed. By 1952 this was probably around twice the amount consumed during the Victorian era.[11] Yet the quality of coal was poor. During the early 1950s, coal was still rationed for homeowners. This was not due to scarcity, but to allow the highest quality coal to be exported, as the UK focused on paying its Second World War debts. However, poor quality coal termed "nutty slack" was exempt from restrictions. This was a mixture of small chunks of coal (nuts) and coal dust (slack). Questions about the exceptionally low quality of nutty slack that winter were raised in the Commons in early February 1953, but this was after 170,000 tons had been sold to help families eke out their supplies of rationed house coal.[12]

Initially the British government tried to view the smog as a natural disaster beyond its control, but pressure mounted as it emerged that the smog deaths were even greater than those in the cholera epidemic of 1866. In January 1953, Minister of Health Iain Macleod remarked, "Really, you know, anyone would think fog had only started in London since I became Minister."[13] It was July 1953 before the government finally bowed to public pressure and appointed the industrialist and engineer Sir Hugh Beaver to chair an inquiry.

Unlike Lord Newton, who led the 1921 air pollution inquiry, Beaver was not a parliamentarian. He was undoubtedly someone who got things done. His original career intention had been to join the Indian civil service (then part of the British Empire), but on a visit to London he took a job as an assistant in an engineering firm. Despite a lack of engineering training, he quickly grasped technical detail and economic possibilities and went on to learn about engineering, mining and quarrying, and transportation and hydroelectric schemes. He headed a review of Canadian ports; rebuilt the harbor at St. John, New Brunswick, after it was destroyed by fire; designed and built several factories in the UK; and headed the Ministry of Works during the Second World War. After the war he became a legendary figure, heading the Guinness company and several government advisory boards. His legacy can be found in the creation of Milton Keynes and other new towns around London, and more oddly in the founding of the *Guinness Book of World Records*[14]—the result, it is said, of an argument during a shooting party, when Beaver could find no reference book to determine if the golden plover was Europe's fastest game bird.[15] At a time when Roger Bannister had just run the first four-minute mile, Beaver's first record books were provided free in pubs to help Guinness drinkers settle the types of discussion that often happen after a few pints. It is now a world best-seller.

In 1955, shortly after the publication of his report on the London smog, Sir Hugh presented his committee's work at the first International Congress on Air Pollution, held in New York. His explanation was calm and authoritative. He told the audience how the deaths of so many in 1952 had caused public outcry and a return to fogs the following winter had led to a fear of air pollution for the first time. He also explained how the publication of his committee's report at the start of the 1954 fog season had built a clamor for action the likes of which had never been seen before. The report's conclusions ran to just 12,000 words. National rather than local controls on industry were proposed. This would prevent towns and cities turning a blind eye to pollution laws in case they threatened local jobs. He found that around half of the UK population lived in areas where action was needed. Importantly, for the first time, Beaver proposed to address the air pollution from home fires, which consumed around 20 percent of the UK's coal but produced around 40 to 60 percent of the nation's smoke. He proposed that clean air or smokeless zones should be set up, restricting the fuels that could be used, and also restricting certain types of fires and boilers.[16]

The aftermath of 1952 not only proved that smog was a killer, but also began to change the scientific view on everyday exposure to air pollution. Addressing the Royal Sanitary Institute, E. T. Wilkins said, "The effects of normal pollution undoubtedly represent a greater damage and loss to individuals than the occasional smog." It would not be until the 1990s that this became the consensus view.[17]

Beaver thought that public opinion now favored action, but government again sat on its hands. A cheery hearth was the heart of the British home. Soldiers in the First World War had been urged to fight to keep the home fires burning. Despite the attempts of smoke abatement societies, the UK was wedded to the open fire. Closed stoves had long been dismissed as a continental eccentricity, making rooms stuffy and unsuitable for the English, who preferred fresh air to warmth, despite the greater efficiency and lower smoke output that stoves offered. At the start of the twentieth century, moreover, central heating was still deeply unpopular. Even offices were heated by open fires. Smoke abatement campaigners recognized how difficult it would be to convince ministers and civil servants to take action, commenting that "a government clerk finds he is terribly injured if he has not a fire in the grate to poke."[18] In 1948, 98 percent of UK living rooms had an open fire and a quarter of homes still cooked with coal.[19]

When the MP Gerald Nabarro[20] began the process of introducing his own clean air act into Parliament, the government was shamed into action and put forward a clean air act of its own. Although hailed as an environmental hero by some, Nabarro was a controversial character. A very well-known politician at the time, he had the appearance of a Conservative "toff," with a distinctive handlebar moustache and booming baritone voice. Despite his aristocratic persona he came from a humble background and was the archetypal self-made man, having gone from yard-laborer to tycoon. He was responsible for many things that we would consider valuable today, including the addition of health warnings onto cigarette packets, but his views on race, capital punishment and a host of other matters were viewed as objectionable and even offensive in the context of the 1950s and 1960s. He opposed Europe, the abolition of capital punishment, drugs, students, pornography and pop music. He supported white rule in Rhodesia and Enoch Powell's opposition to immigration. His downfall came in 1961 when his car, with the famous number plate NAB 1, was seen speeding round a roundabout in the wrong direction. He maintained that his assistant was driving, and she was happy to take the conviction, but witnesses disagreed and the court cases that followed ended his career.

One of the greatest innovations of the Clean Air Act was the smokeless zone, where only approved fuels could be burnt in appropriately approved appliances. Poor-quality (bituminous) coals were banned in domestic open grates and instead only manufactured smokeless fuels or clean coals were permitted. A responsibility was placed on the householder and also on the coal merchant, who had to ensure that only the correct fuel was delivered. Importantly, generous funds were provided to help people to upgrade their home heating.

Many hail the 1956 Clean Air Act as a great success, but progress on smoke control areas was slow. By 1963 only 14 percent of the areas identified by the Beaver Committee were subject to smoke control. By 1967 a disappointing two-thirds of London was covered. In conjunction with increased controls on industry, it was perhaps the availability of new fuels and heating systems that finally made a difference. These included electric overnight-storage heaters, oil heating for buildings and, from 1967, natural gas.

When the Clean Air Act passed into law in 1956, no one knew that vast gas supplies lay under the North Sea and so it was not part of the cleanup plan. Initially natural gas threatened to throw the act into disarray. Smokeless coals were a byproduct of the manufacture of so-called town gas, originally used for lighting in Victorian times, and then for cooking and occasional heating. Every town had its own gas works, creating smokeless coal as a byproduct. In 1970, with gas works closing, shortages of smokeless coal caused large problems for homeowners, delaying the already slow roll-out of new smokeless zones. The problem was simply one of transition. Rapidly, gas heating outstripped coal. Ripping out fireplaces then became popular among do-it-yourself home modernizers and the television replaced the fire as the focal point of the living room.[21]

Table 1 Notable UK smogs

Location	Year	Month	Duration (days)	Additional deaths
London	1873	Dec	3	270–700
London	1880	Jan	4	700–1,100
London	1882	Feb	N/A	N/A
London	1891	Dec	N/A	N/A
London	1892	Dec	3	~1,000
Glasgow	1909	Nov	8	N/A
Glasgow	1909	Dec	4	N/A
London	1921	Nov	5	Not statistically significant
Glasgow	1925	Nov	7	More than 200
London	1935	Dec	6	~500 but very cold
London	1948	Nov	6	~300
London	1952	Dec	5	4,000 (12,000 from reanalysis)
London	1956	Jan	4	800–1,000
London	1957	Dec	N/A	300–800
London	1962	Dec	4	340–700
London	1975	Dec	3	Not statistically significant
London	1982	Nov	N/A	N/A
London	1991	Dec	4	101–178
UK-wide	2003	Aug	N/A	423–768
UK-wide	2014	Mar and Apr	Series of episodes	1,649

N/A = not available

With the health impacts of smog now recognized, reanalysis of mortality records showed the huge death toll that London's smog had inflicted before 1952 and how the 1952 smog was not the last. Table 1 shows data from major UK smogs.[22] In 1954, the reanalysis revealed death tolls totaling nearly 1,000 during the major smogs of the Victorian age. Neither were smogs confined to London. Notable smogs also occurred in Glasgow in 1909, when death rates increased in the young and elderly, and in 1925 with more than two hundred extra deaths. It also highlighted that the important warning sign of around three hundred extra deaths in a thick smog in 1948 had been ignored.

The last great coal-induced London smog was in December 1962. By the late 1970s home coal-burning had been consigned to history, but it was replaced by air pollution from traffic pollutants. London's first major traffic pollution smog occurred in London in 1991, leading to the early deaths of around 101 to 178 people.[23] London and other cities in the UK have had smogs since 1991, but the focus of health research has moved away from quantifying their impacts. However, short periods of high air pollution can still have serious effects on the population, as shown in Table 1.*

The tragedy of the 1952 London smog showed for the first time that air pollution was harmful, ending a debate that had been taking place for centuries. The policies that followed no doubt saved the lives of many, but more could have been saved if the early warnings had been heeded or simply if the government had been

* These last two assessments were health impact analyses: calculation of the likely impacts based on the concentrations of pollutants and our knowledge of the harm that they cause. The earlier ones had looked at death certificates and death rates.

prepared to believe that air pollution was harmful. Looking at the causes of deaths in the 1952 smog, it was clear that air pollution did more than cause fatal breathing problems; around 21 percent of deaths were due to heart problems and around 5 percent were due to strokes, important information that was overlooked until the 1990s.

The 1952 smog did create two wrong directions that still echo in the public and political understanding of air pollution. First, that air pollution is only harmful when it is really bad, and that therefore control of smog events would be sufficient to control the health hazard. We still see this emphasis today, in the reporting of Beijing and Delhi smogs for instance. Second, the conclusion that air pollution is mainly a local problem: the belief that damage to residents' health comes largely from sources within the city they inhabit. As we shall see, these short-term and local perspectives of air pollution ignore the many wider and long-term impacts.

The 1952 smog changed the way that we control air pollution, evolving from looking at factories and the local harm that they cause on a piecemeal basis, to national controls on industry and the management of air pollution area by area. With the benefit of hindsight, however, the solutions offered by the Clean Air Act sowed the seeds of the next air pollution crisis. The Beaver report and the Clean Air Acts that followed focused on smoke but not on the sulfur dioxide that also came from burning coal. Burning smokeless coals did nothing to reduce this pollutant. Londoners were breathing more sulfur dioxide in the 1962 smog than they breathed during the 1952 incident. Allowing sulfur emissions to continue without controls contributed to the acid rain and forest dieback events that dominated the air pollution debate in the 1970s and 1980s and still cause problems today (see Chapter 6). The solutions to London's smogs did not focus on reducing consumption. Improving efficiency and reducing the amount of coal burnt would have lessened the impacts of what we now know as climate change, but this was not realized in the 1950s.

The eventual solution to the UK's problem of urban coal-burning came from natural gas. This was so successful that it not only displaced coal as a heating fuel but oil heating too. As we shall see in Chapter 10, this led to Europe's twenty-first century crisis with diesel car exhaust and the Volkswagen emissions scandal that came to the fore in 2015.

A simple statue in the Belgium town of Engis commemorates the victims of the Meuse Valley smog. Reflecting the industrial nature of the smog, the final line of the commemorative plaque reads, "all human endeavours, even industrial ones, are perfectable." The victims of the Donora smog are remembered in a museum in the town and by a group of locals who are determined that the lessons from the event are not forgotten. Yet look around London and you will find no memorial to the 12,000 people who died from the 1952 smog, although death rates exceeded those during even the worst nights of the Blitz. On January 31, 1953, just seven weeks after the London smog, an unprecedented storm surge hit the North Sea coasts of Europe, causing flooding that killed 326 in the UK and 1,800 in the Netherlands. Travel around the east coast of England from Lincolnshire through East Anglia to Canvey Island and you can find simple, fitting memorials to mark the neighbors, friends and family who died. As London struggles with modern air pollution, it would be fitting to create a permanent memorial to the thousands who died in the impenetrable smog of 1952, to remember those who perished in the smogs that preceded and followed it and to signpost a warning for the future.

Chapter 4

The madness of lead in gasoline

In the 1920s, humanity entered into a Faustian pact, selling our future health and environment for short-term gain—all thanks to Thomas Midgley. He was a prolific US chemist and inventor whose inventions were at the core of two of the most transformative devices of the twentieth century: the lead additive that was added to gasoline and the Freon that made safe fridges and air conditioning possible. At the time of his death in 1944, Midgley was regarded as one of the greatest inventors of the age, but the use of his innovations had massive downsides. By the end of the century, his epitaph had changed as lead became recognized as a global toxic pollutant and Freon was found to have been instrumental in harming the stratospheric ozone layer. He is credited with having "had more impact on the atmosphere than any other single organism in earth history."[1] This book about the harmful air pollution that we breathe would be incomplete without the story of the gasoline additive tetraethyl lead (TEL).

After working for his father's tire development company, Midgley moved to work for Charles Kettering, the inventor of the electric starter motor for cars. In 1916, the twenty-seven-year-old Midgley was set to work on a solution to the problem of car engine "knock." Depending on the severity of the problem, this can be heard as relatively harmless "pinking" sounds, when an engine is under load, or bangs from early detonation that can destroy an engine. Knock limits the efficiency, the power and sometimes the lifetime of an engine. It probably led to early automobiles being dubbed "old bangers." Midgley came up with 143 fuel additives to deal with the problem. The initial front-runner was ethyl alcohol, made from grain and crop wastes, but there was little money to be made from this; anyone could manufacture it. Midgley's final chosen solution was a lead compound first discovered in Germany that could be patented as an additive and manufactured at substantial profit.[2] In 1925, sales of General Motors cars had fallen behind Ford. At the forefront of the fightback was the high-performance Cadillac, but its engine had severe knock. The solution was Midgley's additive.

The neurotoxic properties of lead were well-known. In the first century AD, a Roman physician noted that "lead makes the mind give way." At that time exposure came from the leaded glaze on pots and the use of lead in wine making, where lead plates were dipped into the barrels during fermentation. The latter formed lead acetate or so-called sugar of lead. The same sweet taste encouraged children to chew the lead paint on their toys and cots because it tasted like lemon drops. In the eighteenth century, severe outbreaks of lead poisoning occurred each autumn in Devon and lead in wine making caused epidemics of lead colic. A ban on the use of lead coils in rum making was one of the first public health laws in the Americas in 1723. But, despite this knowledge, poisoning continued. In 1818, while he was ambassador to France, Benjamin Franklin saw the stomach aches and wrist "drop," a type of paralysis from lead poisoning due to lifetime exposures in certain professions. Franklin noted that the "mischievous effect" from lead had been known for at least sixty years and asked "how long a useful truth may be known and exist before it is generally received and practiced upon." In the case of lead it often took many years for the effects of chronic exposure to manifest.[3]

With the toxic effects of lead well-known in the 1920s, Midgley had some work to do to convince the government that his TEL additive was safe for use. An early decision was made to market the product as ethyl, with no mention of the word lead. Three of the United States' largest companies, DuPont, Standard Oil and General Motors, got together to create the Ethyl Corporation. General Motors paid the US Bureau of Mines to investigate the product, under rigid control, which included the strict replacement of the word "lead" by "ethyl" throughout the project. One dissenting scientist who questioned the independence of the report found that their long-standing contract was not renewed, a pattern that would become familiar in the years to come. While the Bureau of Mines found no evidence of harm, production of TEL began with disastrous consequences for some of the employees.

On October 26, 1924, Ernest Oelgart, a worker at Standard Oil's TEL laboratory, became paranoid, running around the plant saying that "three were coming at me." He died shortly after in a hospital. Over the next five days, four more workers died and thirty-five showed neurological symptoms of lead poisoning. Six workers died at two other plants, and one TEL plant was nicknamed the "House of Butterflies" as workers hallucinated about flying insects. However, tests on the excreta of workers at the plant showed no difference between those working with TEL and those elsewhere in the plant. Company tests found lead was in the excreta of Mexicans as well as the people at the TEL plants, allowing them to conclude that the presence of lead in the human body was normal. The Ethyl Corporation asserted that these workplace incidents did not imply risk to the public; the workers were either careless or had simply been working too hard. At a one-day science event that followed, the president of the corporation put the burden of proof back on the public health scientists. He challenged them to show that TEL, "a gift from God," was dangerous because the gasoline-saving engine improvement was certain. Midgley was so confident in the use of his product that he poured it on his hands and took deep sniffs in front the of press, despite having only just returned from time away from the plant to recover from an earlier bout of lead poisoning himself. He boldly declared that the use of TEL would be undetectable in the air beside a busy road. A short trial at 252 gas stations revealed no impacts, and TEL was approved for use in 1926.[4]

TEL use grew, and at its peak some 200,000 tons of lead was being expelled into the air from European and US vehicles. For the next forty years, all studies on TEL were funded by the chemical and auto-and-oil industries under the umbrella of the toxicologist Robert Kehoe, who had investigated the deaths of the TEL workers. From his research, the product retained a clean bill of health.

It took someone completely outside the field to finally question the established view. Clair Patterson—"Pat" to his friends—was a geologist who had been looking at metal isotopes in rocks. Patterson was born in 1922 in Iowa. His father was a postal worker and his mother served on the local school board, but it was a chemistry set given to him as a child that first piqued his interest in science. By the 1960s he had made his name by working out that the earth was much older than previously thought. He then set about investigating how the earth had changed over the last 4.5 billion years. In 1965, he published a strange finding that had emerged in his samples from the Atlantic and Pacific Oceans; the amount of lead currently entering the oceans each year was around eighty times greater than the historic rate. The waters at the top of the ocean contained up to ten times the amount of lead found in the depths. Importantly, Patterson showed that the normal concentrations were far from natural. He estimated that the lead in the blood of the average American might be 100 times the natural level and approaching the levels where toxicological effects had been seen.[5] This challenged the industry-led consensus, and Kehoe was not happy. He accused Patterson of rabble-rousing and of being more of a zealot than a scientist. Patterson recounts that a group from the Ethyl Corporation visited him and, in his words, "tried to buy me out through research support that would yield results favorable to their cause." He refused. He was publicly vilified and his professionalism questioned. His long-standing research contracts were terminated, and a delegation visited his department head to get him fired. Patterson persisted and went on to find that lead from TEL had spread throughout the world. The lead in Greenland ice was one hundred times greater than pre-industrial levels and snow in remote Antarctica contained ten times more. Later he would be awarded the Goldschmidt Medal, the equivalent of a Nobel Prize in geochemistry, elected to the National Academy of Sciences and have both an asteroid and a mountain peak named after him.

Although Patterson showed that lead from gasoline had become a global pollutant, he was not a toxicologist. Several investigations followed but these were heavily influenced by the involvement of the TEL industry. Then along came Herbert Needleman. In the early 1970s he was working at a community psychiatric clinic in north Philadelphia when he met a young man with big ambitions, a bright lad but one who struggled with his words. His problems reminded Needleman of children with lead poisoning and he wondered if lead was a factor in the problems of some other patients. It was hard to know how much lead was present in children's bodies. Tests on urine and excreta only gave information on recent lead exposure and did not shed much light on the amount that had accumulated in their bodies. This had been a serious flaw in Kehoe's earlier examination of the TEL workers.

Needleman's office overlooked a children's playground, which gave him an idea. Needleman became the tooth fairy. Over two thousand children were paid for their milk teeth, their teacher verifying every child's new socket to be sure that the tooth had been theirs. Each tooth was analyzed for accumulated lead. Needleman asked the teachers to grade the children's behavior and the children were also tested at a clinic. Those with high amounts of lead in their bodies tended to perform worse on intelligence tests and at repeating sentences and rhythm; they had slower reaction times and displayed hyperactivity. Children from busy urban neighborhoods had higher levels of lead than those in the suburbs. It was clear to Needleman that something needed to be done about the lead that children were exposed to in their everyday lives.[6] But not everyone agreed.

The auto and oil industry responded with unprecedented opposition and set out to discredit Needleman. Two young scientists came to visit him, saying that they had an interest in lead exposure near a smelter. To help them, Needleman shared his data. Apparent irregularities were discovered and, following a familiar pattern, Needleman found himself accused of scientific misconduct. A case was taken to the newly formed federal Office for Scientific Integrity and this was followed by an inquiry at his own university. Officials came into his office and screwed locks to his filing cabinets and drawers. Years of uncertainty ensued, but Needleman remained resolute and was eventually cleared of any wrongdoing.[7]

Despite the growing weight of evidence of the environmental and health harm caused by TEL, the end to leaded gasoline was not brought about by scientific findings or by public outcry, but by the advent of the catalytic converter. These were introduced in the United States in 1970 to clean up other pollutants from gasoline engines and to tackle the Los Angeles smogs (see Chapter 5). However, platinum in the catalysts was poisoned by leaded gasoline, so the leading had to stop. As Needleman and colleagues remarked, "Apparently, poisoning a technology was more important than poisoning people."[8]

In 1970s Europe, scientific recommendations resulted in reductions in the lead content of gasoline, but public pressure against lead really began to mount in the 1980s. In the UK, property developer Godfrey Bradman funded a national campaign to have lead removed from gasoline. He recruited the seasoned political campaigner Des Wilson to run the Campaign for Lead Free Air (CLEAR).[9] Wilson was a journalist who had come to the UK from New Zealand in the 1960s and was a breath of fresh air in stuffy British public life. Before his work on lead-free gasoline he had already founded the homelessness charity Shelter, creating a new form of campaigning charity that had not been seen in the UK before. In 1983, two years into Wilson's campaigning against lead, the Royal Commission on Environmental Pollution recommended that lead in gasoline be reduced. Government accepted the findings within thirty minutes. It was a triumph for Wilson and the scientists and activists who worked with him. The campaign was ended, its job was done, and Wilson went on to chair the UK's Liberal Party and head Friends of the Earth.

Sadly, the Royal Commission's second recommendation, that lead be removed from gasoline completely, was not implemented until sixteen years later, in 1999. The UK waited until the latest time allowable under EU law before it banned lead in gasoline, thirty years after the phaseout began in the United States. It was a betrayal of Wilson and those campaigning for lead-free gasoline, who believed that the argument had been won. The UK's reluctance to phase out leaded gasoline completely was most likely for two reasons: UK car manufacturers were slow to invest in engines that could run on lead-free gasoline and a factory in the UK was Europe's largest manufacturer of TEL. Ten years after the UK ban was announced, the Octel plant, alongside the Manchester Ship Canal, was the last place in the world still making TEL. After the UK, European, and US bans, Octel had created new markets for lead gasoline additives in the developing world, generating new sales of $1.8 billion and profits of $600 million. In 2010, Octel was found guilty in UK courts of bribing the head of the Indonesian state oil company, financing "Defence of lead" campaigns and delaying a ban through "a slush fund to corrupt government figures with the intention of blocking legislative moves to ban TEL."[10]

Taking lead out of gasoline worked. As it was removed, the blood lead levels of children fell, but it would be complacent to view this as a problem solved. We are all exposed to lead in our everyday lives, we all have it deposited in our bones and circulating in our blood. In the mid-1990s, well after lead in gasoline had been banned, blood samples were taken from 14,300 American adults. Researchers visited them all again in 2011. Those with the highest blood lead levels were shown to be dying earlier from heart disease, and the risks could be seen even in those with low concentrations of lead in their blood. The battle against toxic lead additives in gasoline might be largely won, but more action needs to be taken on our other exposure routes, including our drinking water and food.[11]

Important lessons can be learned from the lead story. Crucially, no evidence of harm is not the same as evidence of no harm. There is also a difference between what the industry presented as a normal level of lead in our bodies and what should have been there naturally; these are not the same when masses of people are being exposed to a toxin. During the debate on the safety of TEL, the onus was shifted onto public health specialists to prove that TEL was harmful. It should have been the responsibility of the Ethyl Corporation to prove that it was safe. Early warnings were ignored by an industry that was keen to create a market for their new product. Any new research that damaged their business was systematically suppressed and the researchers discredited. Many people were bought off or given funding or jobs to support the industry. It seems that lead not only affected the minds of children around the world but also the morals of those who made TEL, and the consequences can still be seen today.

Chapter 5

Ozone, the pollutant that rots rubber

Today, when we think of ozone, most people think of the so-called ozone hole in the stratosphere. However, as the early explorers of our air discovered, ozone is also present in the air that we breathe at ground level. The gas is the same, but the problems with stratospheric ozone and ground-level ozone are very different.

Ozone in the stratosphere is considered to be a good thing, protecting us from the sun's harmful ultraviolet radiation. In 1985, scientists working for the British Antarctic Survey confirmed that the ozone layer was thinning. Suddenly we learned that our aerosol deodorants and the fridges that we had taken to the waste dump had damaged our global environment. The boycotts of aerosols and the public clamor resulted in the 1987 Montreal Protocol, which controlled the production and use of ozone-depleting substances. This agreement is a unique example of international action to protect our atmosphere, not least for the speed it was put in place. The protocol also set a framework for the developed and developing world to work together to provide support and fund its implementation.

International control of the ozone that we breathe at ground level has been much less successful. The current and future health impacts of ground-level ozone and the damage to our food crops should concern us all. It can only be resolved by concerted global action, which sadly seems a distant prospect. The story of our understanding of this pollutant matches the familiar pattern of initial denial of the problem by industry and government, until confronted by a sudden new discovery, and then pressure from the public as they realized the global scale of its impact.

On July 26, 1943, at the height of the Second World War, residents of Los Angeles thought that they were under chemical gas attack.[1] People in the downtown area experienced stinging eyes, streaming noses and rasping throats. Many office workers had to be sent home from work. A haze had descended on the city, reducing visibility to less than three blocks. It lasted for days. Photographs of the event look nothing like the smoky smogs of London and eastern US cities, and in any case the smog was in the middle of a heat wave. Distressed people turned to the city's authorities for help, but no one had a clear answer about where the smog came from or when it was going to go away. Increased traffic due to a public transportation strike was an early explanation, but this was rapidly dismissed. The *Los Angeles Times* reported that children were unable to concentrate with stinging, smarting eyes, and in one case a child's eyes were swollen closed. The paper championed the need to clean the air and brought in experts to write about possible solutions. Reporters described the smogs as "visibility reducing palls and eye tearing sieges."[2] Over the previous decades people had moved away from the industrial east to Los Angeles precisely because of its clean air and healthy lifestyle, but something had clearly gone wrong.

Fingers were first pointed at a synthetic rubber factory, which had just gone into production for the war effort, but smog both upwind and downwind of the factory meant that it could not be the cause. Initial actions focused on solutions imported from other cities, mainly related to reducing smoke. Some were useful, such as a ban on back garden rubbish-burning and setting up a mandatory rubbish collection, but it was clear that the LA smog was something not seen elsewhere and it required new solutions.

It fell to the chemist Arie Jan Haagen-Smit to explain and solve the problem.* Haagen-Smit, or "Haggy" to his friends, was a plant scientist and biochemist. Born in the Netherlands, his father was a chemist at the Dutch mint. Having toyed with the idea of becoming a mathematician, Haagen-Smit decided that his work prospects would be better if he followed his father and became a chemist. As the specter of war loomed over Europe in 1936, he was recruited to the United States on a one-year contract but quickly established himself within the California Institute of Technology. At the time of the Los Angeles smog he was busy isolating the chemical that gives pineapple its flavor. Haagen-Smit turned briefly to look at the smog problem following the 1943 "gas attack." His initial investigation pointed the finger at the petrochemical industry and he returned to his pineapples. That would have been the end of his role in air pollution science if industry had not hit back. Haagen-Smit's wife, Zus, recalls Haggy's anger and hurt when his work on smog was

* To learn more about the man, have a look at http://calteches.library.caltech.edu/368/1/haagensmit.pdf.

publicly criticized. One vocal critic from the Stamford Institute of Technology, who was funded by the oil industry, wrote a note to Haagen-Smit saying, "You know where my livelihood comes from. You know, I have to say these things."[3] Haagen-Smit reluctantly set aside his food flavoring work to rebut the criticism and reestablish his reputation by finding the cause of the Los Angeles smogs.

With his background in plant biology it is perhaps not surprising that he used plants in his investigation. A huge sealed Plexiglas chamber was built in a car park, open to the sunlight so that Haagen-Smit could mimic the smog. Knowing that the new type of smog damaged crops, he took spinach, beets, endive, oats and alfalfa and exposed them to vapors from distilled gasoline and nitrogen dioxide (from traffic exhaust), ozone and ultraviolet light, to find out what the harmful agent was. Similar experiments were carried out with people who were sensitive to smog. Volunteers were shut in the chamber and scientists timed how long it took until tears ran down their faces. At times a thick blue haze was created in the laboratory, making it impossible to see more than a few meters. Slowly the complexities of the smog were unraveled.

Measuring the smog outdoors was harder. Like John Switzer Owens, Haagen-Smit needed an easy way to take measurements all over the city. This ruled out a lot of specialist laboratory-based work. Instead he looked at the damage the smog was causing. The LA smog attacked items made of rubber, causing them to crack. Before the widespread use of plastics, rubber had many more uses than it does today, and for rubber to crack was a big problem. Cleverly, Haagen-Smit's team simply placed stretched rubber tubes outside each hour and waited until cracks appeared. Sometimes it took forty-five minutes or more for the rubber to start to crack, but during smogs it took as little as six minutes.

Finally, Haagen-Smit solved the mystery. The 1943 "gas attack" and the smogs that followed were not from burning coal or waste but were being created in the Los Angeles air. Sunlight was interacting with the air pollution and the city behaved like a huge chemical reaction chamber. The resulting smog was created from the traffic exhaust, including nitrogen oxides and unburnt fuel vapors, and the evaporation of petroleum from refineries and filling stations. Haagen-Smit had discovered and understood a completely new type of air pollution that was coming from a new and growing source: the motor car and its fuel. In 1952, just a few months before the great London smog, he published a seminal study about the Los Angeles smog and ozone pollution.[4]

Solving the problem was not going to be easy in an area with a powerful oil industry and a growing car-dependent population. The car was a symbol of progress and part of the American dream. Unsurprisingly, the early fightbacks came from the refining industry, who maintained that ground-level ozone was natural, brought down from the stratosphere and in no way connected to petrochemicals. However, this argument was soundly defeated in 1954 when measurements revealed that Catalina Island, just off the California coast, had very little ozone, while the residents of LA were in enveloped in smog. The problem was localized in and around the city.

According to Haagen-Smit's wife, the oil and automotive industries fought back tooth and nail every step of the way.[5] Haggy had the respect of the engineers and scientists, but the senior management in the oil and automotive industry resisted anything that would increase costs. Nevertheless, regulations were slowly put in place. The first thing to be controlled was the 700 tons of gasoline evaporating from refineries and filling stations each day. New roofs on storage tanks more than halved this amount, and controls at filling stations trapped vapors too. This was followed by laws in the 1960s to improve fuels, including the removal of olefins, some of the most potent ozone-forming chemicals. However, it was not until the 1970s, when the first catalytic converters went on the market, that real advances were made in cleaning traffic exhaust. As governor of California in the late 1960s, Ronald Reagan played a role in the battle. Although not famed for his environmental actions as US president, Reagan created the California Air Resources Board (CARB). Ever since then, CARB has been a world leader in the control of air pollution, most especially from traffic, and played an important role in uncovering the Dieselgate scandal (see Chapter 10).

When he died in 1977, Haagen-Smit was one of the most famous environmental activists in the United States. But first and foremost he loved the science of smog; how it changed each day and where it came from: "If I look at the daily graph to see how much smog is in the air," he once admitted, "I feel a twinge of disappointment when there's just a small amount."[6]

Los Angeles–type smog started to be found elsewhere—in other hot places in the Americas, such as Mexico City—but any notion that Europe could experience a Los Angeles smog was dismissed with confidence by the Royal College of Physicians. Strong sunlight was thought to be key to the LA smog and under gray, damp European skies, ozone was not something to worry about. The agenda had been set in the 1950s and the fight for clean air was all about controlling smoke from coal.[7] The medics at the Royal College were not alone in this assumption, which simply reflected the scientific consensus at the time. The measurements from the Observatoire de Montsouris from the late 1800s (see Chapter 2) had shown that ozone formed in the summertime in northern Europe even before the advent of pollution from cars and oil refining, but this data lay forgotten, gathering dust in the Paris archives.

Then, in 1972, a paper appeared in the journal *Nature* with the headline "Photochemical pollution of the kind which occurs in some urban environments has been observed in Britain." Acknowledging that "it is generally held that photochemical air pollution is unlikely to be encountered in Western Europe because of less sunlight," the paper went on to reveal that ozone was being produced on a hot July day in the Oxfordshire countryside. Some patchy evidence from researchers in the Netherlands and Germany suggested that the Oxfordshire event might not be unique. In just thirty-five days of measurements, US health standards were exceeded five times in the green English countryside.[8] Maybe we had been too complacent, or perhaps too focused on coal and smoke? The more researchers looked, the more ozone-filled summertime smog they found.

These findings emerged at a time when the UK government's air pollution laboratory was coming under criticism that it needed to improve the quality of its science.[9] The response was a more ambitious research program. A wider team of scientists was assembled, and a line of measurement sites was installed stretching from eastern England to southern Ireland. They gained access to a water tower in East Anglia and enlisted help from local councils in Oxfordshire and Adrigole, near Cork.[10] Amongst the scientists was

James Lovelock, who was just about to become a household name. His renown would not be for air pollution research, or for his inventions, but as the environmental free-thinker who first put forward the Gaia hypothesis, which opened the minds of scientists to new ways of thinking about our planet. It captured the imagination of a generation in the 1970s when environmental activism was emerging and it is still debated today.* During the early 2000s, Lovelock would wrong-foot and disappoint many environmentalists by advocating nuclear power. He argued that this was a necessary solution to combat climate change and that humanity needed to adopt a new defensive position to address the inevitable impacts of the earth's future climate, including the construction of new ports and flood defenses. Another member of the UK team was Richard "Dick" Derwent,[11] a recent Cambridge graduate who would be a constant voice in air pollution research for decades to come. An atmospheric chemist with an insatiable appetite for data and an encyclopedic knowledge of monitoring stations, Derwent was an enthusiastic supporter of everyone (like me) who measured the composition of our atmosphere. He went on to explain the air pollution in London's first traffic smog of 1991 and the way that air pollution behaved across Europe and the globe.

The team found that the earlier measurements in the Oxfordshire countryside were not a one-off. Yet again they found that health standards, which had been devised to protect people in the United States, were being breached over southern England, but the sheer

* Put briefly, the Gaia hypothesis says that life on earth engineers and creates a sustainable environment for itself in a self-regulating system that has kept our planet habitable for the last two million years. See https://www.newscientist.com/round-up/gaia/.

area covered by the smog was a revelation. Ozone and other smog pollutants could travel over 600 miles. Haagen-Smit's view of the ozone in Los Angeles led to the idea that it was a city-region problem, but the UK team showed that this pollutant was not going to be tackled if each town worked on its own. More measurements followed, revealing that US standards were being breached in London; the air pollution in London was not so different from that in LA after all. Ozone was especially severe during the 1976 heat wave due to pollution from sources outside the UK. Not only did ozone form over large distances, the time it took to form in the air was affecting which day of the week was the most polluted; days at the end of the working week had higher levels of ozone compared with Mondays and Tuesdays, which benefited from the weekend's lower traffic volumes.[12]

Ozone over the UK was found to be the least of Western Europe's ozone problems; the worst-affected areas were found to be in the Mediterranean. The Po Valley in northern Italy emerged as a specific hotspot. Like LA, this is a basin where polluted air frequently becomes trapped, in this case between the Alps and the Apennines in an area that is home to twelve million people and the location of massive industry around Turin and Milan. It became clear that Europe had a lot of catching up to do to get this pollutant under control, but unlike the challenges faced by LA, the solutions were already there.

Great progress was made in the UK from the 1970s, mainly in controlling pollution from traffic and the oil industry, but—as we have seen—the introduction of catalytic converters to vehicle exhausts was delayed until the 1990s, partially due to the continued use of lead in Europe's gasoline. Ozone levels during UK heat waves since 2000 have been around one-third of those in the heat wave of 1976, but this pollutant was still thought to be responsible for between 423 and 769 of the extra UK deaths during the record-breaking temperatures of 2003.[13] The Los Angeles experience told us that we should be worried about smog episodes. More information came from studying adults exercising in pollution-filled chambers and from children at summer camps who suffered breathing problems on smoggy afternoons. However, in the last decade we have learned that the ozone we breathe every day could be shortening our lives.[14] More worrying was a study published in 2017 that used health records from the sixty-one million elderly Americans who were part of the Medicare system.[15] This found that lives were being shortened by ozone and in some cases, this was happening to people who lived in areas that met current US standards. Current standards need to be tightened and decreasing ozone still further will require new actions.

Chapter 6

Acid rain and the particles that form in our air

As we left the UK in Chapter 3, smoke from coal-burning was being tackled. The heavy, pea-souper smogs and the dense smoke clouds that hung low in winter skies had largely been condemned to history, but the Clean Air Act only dealt with the smoke from coal-burning. The sulfur from burning coal was left untouched. This sowed the seeds for the major environmental crisis of the 1970s and 1980s. It started in a small way with the concerns expressed by a small number of ecologists, but went on to cause major international tensions in Cold War Europe[1] and culminated in one of the greatest air pollution battles yet. It was not just an argument between the public and governments who were reluctant to act but also between the countries of Europe and North America.

By the 1970s, the postwar expansion of industry meant that Europe was burning more coal than ever before. The solution to the problems of coal smoke had not been to burn less coal. We were burning masses of oil too. Both fuels contain sulfur and, as a result, the amount of sulfur being pumped into Europe's air had nearly doubled in two decades. This had serious consequences.

For a long time it had been known that sulfur gases in our air made our rain acidic. In the 1870s, Robert Angus Smith had found that "rainwater in town districts and even a few miles from towns is not pure water for drinking." For years, the acid in rain had been attacking the stonework of historic buildings and statues. By the 1950s, the carved faces of the statues of generals and politicians of the past were losing ears and noses, gargoyles were looking less ugly and engraved mottos on buildings were fading to become unreadable.

But when acid rain fell in the countryside, Smith found, "the impurities are completely removed by filtration through the soil." It was the prevailing wisdom, even in the 1970s, that the natural environment simply absorbed our pollution and rendered it harmless. It was a happy balance of industry and nature.[2] This could not have been more wrong.

The first alarm bells were raised by a Norwegian scientist called Brynjulf Ottar. He was the first director of the Norwegian Institute for Air Research, founded in 1969, and was also its first employee. Ottar spoke up and spoke up clearly. The rain across Europe was changing.[3] In 1955, acid rain had been confined to central Europe. It stretched as far north as Denmark, but Scandinavia was untouched. Just fifteen years later, the rain that fell over most of Sweden and the southern half of Norway was acidic. The soil types in Scandinavia were especially vulnerable, and the impacts were catastrophic for many forests, rivers and lakes. Acid rain affected over a quarter of

Sweden's 90,000 lakes and 4,000 were rendered lifeless, while previously healthy pine trees were left looking like fence posts. The images of dying trees galvanized European public pressure, especially the green movement in Germany. *Waldsterben*, the German word for forest dieback, captured the zeitgeist and became a buzzword in the Western media. Acid rain damage was also found in the lakes and forests of eastern Canada and the neighboring US states.

The suddenness of many of these events was shocking. A single rain shower could contain nearly a third of a woodland's annual dose of sulfur, quickly overwhelming the ecosystem. The spring melt of acidic snow could kill fish in lakes and across whole river systems.[4]

Everyone thought that these problems were coming from local industry and nearby cities, but this did not make sense to Ottar. He carefully worked out how much harmful sulfur was landing in Scandinavian forests and compared this with the sulfur put into the air by the coal and oil burned in the Scandinavian countries. The amounts did not add up. The harmful sulfur pollution landing in Scandinavian forests, lakes and rivers was much greater than these countries were producing themselves. Additional pollution had to be coming from other European countries. It was not a national problem but an international one, and would require other countries to control their pollution.[5*]

The next step in the escalating international crisis came from the rising tensions between the Warsaw Pact and NATO countries.

* This was not the first time that air pollution had been shown to travel large distances and cause damage to ecosystems. In 1661, John Evelyn reported complaints from French winemakers that their vineyards were damaged by smoke drifting from England.

In 1983, the Russians had just shot down a Korean civilian airliner, mistaking it for a US spy plane. Massive military exercises had taken place in Europe and US president Ronald Reagan's administration was deploying new medium-range nuclear missiles in West Germany. All sides agreed on the need for some sort of détente and the environment was perhaps the least contentious issue for both sides of the Iron Curtain. The USSR and Warsaw Pact countries suddenly became strong supporters of sulfur reduction, championing a 30 percent cut, and were the first to ratify protocols.[6]

Differing views exist about the Russians' motivation. The first view is that the USSR saw positive engagement as a way to appeal to the public in Western Europe and to appear to be the reasonable party being threatened by unwarranted US aggression. The second view is more Machiavellian. Was the USSR's sudden enthusiasm for sulfur controls merely a tool to drive a wedge between the Western allies? If so, it was extremely effective. The blame attached to the movement of air pollution between countries caused tensions between the United States and Canada, between the Scandinavians and the rest of Europe, and between the UK and everyone else.

Ottar's analysis had found that, as a result of industrial coal- and oil-burning and westerly winds, the UK was Europe's largest exporter of sulfuric air pollution.[7] This left the UK isolated in international discussions and labeled "the dirty man of Europe." Moving power generation to the countryside and building tall chimneys had reduced local air pollution but did not prevent sulfur being transported over thousands of miles. It made it worse, and Ottar's team could prove it.[8] They flew aircraft from Norway across the North Sea to the UK, intercepting the drifting sulfur plumes from UK power stations. The idea that the pollution was spread evenly by the tall chimneys was roundly debunked. The plumes did not fan out ever wider with distance, but drifted as horizontal columns until the pollution reached the forests and lakes of Scandinavia. The vast majority of the UK's coal and oil power stations had been built without sulfur controls. There were two notable exceptions: Battersea* and Bankside (now the Tate Modern gallery) in London had sulfur controls, although these were intended to protect nearby historic buildings, not the environment. Both used water washing, a process only possible because the River Thames was so polluted that there were no fish or shore life to be killed by the resulting effluent.[9]

At this time, UK power generation was a nationalized industry. What followed is redolent of climate change denial today; shamefully, UK scientists and civil servants took to the pages of scientific journals and the popular media to discredit Ottar's work. They pointed out uncertainties rather than what was known, creating doubt without attacking the science directly.[10] They also trivialized the forest damage and the impacts of fish dying in acidified Scandinavian rivers. In an editorial in the journal *Nature* with the title "Million dollar problem—billion dollar solution?"[11] they suggested with flippancy and arrogance that rather than clean up its industry, the UK could offer crushed limestone to be sprinkled in Norwegian rivers and, even more incredibly, that the Norwegians would be foolish if they did not accept. The concept that industry in one country should take responsibility for damage in another was also attacked. Instead of incurring the cost of dealing with the

* My dad recalls the white stains left on roads in south London in the 1950s from trucks moving chalk and chalk slurry to Battersea Power Station.

causes, including cleaning emissions or removing sulfur from the heavy fuel oil used in some power stations, other cheaper solutions were put forward, such as covering historic statues in polymer coatings or encouraging anglers to simply relocate from one lake to another when the fish died.[12] Finally, the Thatcher government conceded, and in 1985 the UK signed on to sulfur reduction targets.

Acid rain was not the only problem caused by ignoring sulfur in our air pollution control plans. Sulfur was causing other problems in our cities. By the 1970s, the urban use of coal had disappeared or been replaced by smokeless fuels and natural gas heating. The black smoke that had plagued UK cities was largely gone, but the main method for measuring air pollution was still based on that devised by Owens more than fifty years before and was designed for towns full of coal smoke. After decades of taking them, scientists began to question what these measurements were actually telling them. David Ball and Ron Hume were part of the team looking at this question at the Greater London Council.[13] They uncovered two problems with the conventional view of urban air pollution.

The first problem was that the levels of black smoke pollution in summer and winter had become almost indistinguishable. They no longer fit with the winter heating theory. So what was causing this? The answer for Ball and Hume was easy. At the time, lead was used as an additive in gasoline and Londoners were breathing lead particles every day. When the filters were black, they also contained more lead. So coal was no longer the main challenge for the city: diesel and gasoline exhaust were now the dominant sources of the black particles in London's air. Something had to be done about the traffic.

This was very unwelcome news. Everyone wanted to hear that our air pollution concerns had been tackled.

The second problem was discovered when Ball and Hume compared particle concentration using the blackness scale to the results from the more complex weighing of filters. Traffic and coal- and oil-burning made up only 15 percent. Measuring the blackness of the particles gave them no information at all on the other 85 percent that people were breathing. The vast majority of particles in London's air were simply being missed by the conventional measurement methods.

Back in the 1920s, Owens had noticed haze produced by non-black particles, and the London team now set out to measure this.[14] They illuminated the particles in the air in much the same way that droplets and dust can be seen in a flashlight beam. These particles were very good at scattering light. But what could they be? They were not pollutants that came from any chimney, factory or car. Ball and Hume thought that they might be made in the air from other pollutants: mainly from sulfur gases from burning coal, oil fumes and traffic exhaust.

The late 1980s and 1990s brought about a revolution in air pollution measurement. The new instruments were a huge advance on the British black smoke method. They selected the size of particles to match those we inhale and measured them in real time. Together with new knowledge of the health effects of modern air pollution, UK experts proposed a standard for particle pollution (specifically PM10) and said that it should not be exceeded on more than four days per year.[15] They then set out to see how we were doing against this new health-based standard. The answer was not good. In fact it was really bad.

Smogs had not gone away after all. In March 1996, most of the UK had nearly two weeks of high air pollution and oddly, conditions in the countryside were much the same as in the towns and cities.[16] Was this a return of the particles that Owens had detected, or had they been there all along? The 1996 event was examined by a government advisory committee. One of its members, John Stedman,* remarked, "It is not clear how often we can expect this type of episode to happen again."[17] The answer soon became apparent as new instruments revealed that these previously invisible particles in fact filled European air every spring, consistent with Owens' measurements in the 1920s.

Unlike Owens, however, in 1996 we could measure what the particles were made of, discovering around a third was comprised of sulfate, as Ball and Hume had suspected in the 1970s. Nitrate particles were there too, along with particles of organic carbon. Sulfate forms from the sulfur dioxide created when burning fuels that are rich in sulfur—mainly coal but also oil, diesel and gasoline. Sulfur dioxide is not the only pollutant gas that turns into acidic particles. Nitrogen oxide gases do the same thing, forming nitrate particles. These come from hot flames where nitrogen and oxygen molecules (pairs of atoms) break apart and recombine as a molecule of one nitrogen and one oxygen atom. As I will discuss in Chapter 10, this was the main pollutant involved in the diesel exhaust scandal. The nitrate particles were every bit as harmful to forests and moorlands as the sulfates, and their spread around Europe was studied by Dick Derwent (who we met working alongside James Lovelock).

* We will meet John's uncle in a van in central London in Chapter 10.

All these particles were around the same size as the ones found by Owens during his Norfolk holiday. As Owens discovered, the size of these particles makes them very good at scattering light and reducing visibility. They were almost certainly responsible for the hazy fogs that he investigated downwind of cities and in remote parts of the UK. We were breathing these particles in the 1920s and before, but we did not have the capacity to detect them. They were being formed in the air by chemical reactions between other pollutants. This gave them the ability to cover huge areas and travel large distances. Ball and Hume, and Owens before them, had been proved right after all.

Although we now understand what these particles are, they still plague our towns, cities and countryside, especially in spring. Their emergence each spring catapults air pollution onto the front pages of newspapers across Western Europe. They are a real headache for policymakers trying to control our air pollution. As an emergency response, vehicles bearing odd or even number plates have been banned from the roads in Paris and other French cities on certain days in a vain attempt to keep the pollution under control. In April 2014, Prime Minister David Cameron triggered a media avalanche when he tweeted that he was canceling his morning run due to air pollution.

Although these particles are in our air nearly every day, they appear in much larger quantities during the spring, a fact that seems counterintuitive to our normal image of spring as the time of blossom and fresh new green growth. The answer to the springtime particle problem lies not just within our cities or factories but also in the countryside. The role of agriculture in air pollution will come as a surprise to many, but a crucial ingredient in the secondary particle mix is ammonia. This mixes with sulfur dioxide and nitrogen oxides to change them from gases to form ammonium particles. And the ammonia comes mainly from farms. Across Europe, farm animals dominate ammonia emissions. In the UK, half of these come from cattle manure and slurry and a quarter from poultry. Europe-wide, pigs play an important role, especially in the Netherlands, Denmark and Brittany. In many places, slurry spreading is restricted during autumn and winter to prevent river and water pollution, but this means the practice is carried out during the spring, when fertilizer is also being applied to crops and animals are let out from winter housing. This burst of agricultural ammonia is the cause of the springtime pollution episodes that plague Western Europe each year.

Control is not easy. Reducing sulfur dioxide can lead to fewer sulfate particles but nitrate ones form instead, gobbling up the leftover ammonia. We need to act on all three, especially ammonia. This needs to be done urgently; particle pollution during the spring 2014 episode brought about an estimated 600 early deaths in the UK alone.[18]

Controlling air pollution in Europe and the acid rain problem requires cities and countries to work together, and to extend their scope into farming. Particles can form hundreds of miles away from their source, and, as Owens predicted, they can stay in the air for a week or more because of their small size. To control air pollution in one city, or even on one street, we must therefore take action across a very wide area. It is the same further afield: news stories from Beijing show what is described locally as haze, much as Owens would have seen in Europe in the early twentieth century.

It is interesting to think what would have happened if, instead of British black smoke, the instrument that Owens perfected on his Norfolk holiday had become the standard method of measurement. We might have been able to take a broader view of the particle problem sooner. This could have led to greater control on sulfur emissions, reducing secondary particles and the damage that sulfur and acid rain did to the European landscape.

Despite the strained Cold War politics that surrounded its foundation, the Convention on Long-Range Transboundary Air Pollution is still going strong, decades after it came into force in 1983. Under the umbrella of the United Nations Economic Commission for Europe, it stands as a unique international body working to reduce air pollution across Europe and North America. In successive rounds of agreements, the countries have set targets for future air pollution and worked together to achieve them. It has brought about reductions in air pollution, reduced environmental impacts and strengthened air pollution science. Today, the reduced burning of coal and heavy fuel oils, along with strict Europe-wide laws on chimney emissions, have led to a big reduction in the damage caused by acid rain compared with the 1970s and 1980s. However, 7 percent of the EU still receives more sulfur than its soils can handle. Acid soil and water is still a problem in Scandinavia, parts of the UK and central Europe, and it will take decades for the damaged areas to recover. Ammonium and nitrate air pollution is still causing soil problems across 63 percent of Europe's area, including most of mainland Europe, Ireland and the southern parts of the UK and Scandinavia.[19] Similar problems are becoming evident elsewhere. In 2010, half of China's cities and 40 percent of its land area experienced acid rain. The long-term impacts of this are yet to emerge.[20]

Although our understanding of sulfate and nitrate particles really emerged in the last decades of the twentieth century, our understanding of another category of non-black particles is even less developed. These are organic carbon particles: carbon combined with other elements such as oxygen and hydrogen. We now understand that these particles have a tough, enigmatic time in the atmosphere, reacting with each other and changing from particles to gases and back again to particles with the varying temperatures of the day.* Hydrocarbon particles and gases are attacked and oxidized in the air, in much the same way as a cut apple goes brown.

Some of the organic particles start out as gases from natural sources. A good example is pinene from pine forests, the chemical that gives pine its distinct smell.† Many of them come from burning hydrocarbons, and a huge scientific debate rages about the contribution of gasoline or diesel exhausts to the organic particles in towns, cities and, more controversially, in the air several days downwind.

By working together, the countries of Europe reduced levels of sulfur far beyond the original 30 percent agreed during the Cold War. Along with other Europe-wide pollution controls this was estimated to be saving around 80,000 premature deaths per year by 2011 and saving European economies 1.4 percent of their gross domestic product annually.[21] But is it enough?

* Ammonium nitrate particles do this too. They turn into gases in the heat of a summer's day and reform as particles in the cooler evening. Annoyingly, they can also do this when you are trying to measure them.

† For this reason the planting of some species of tree to improve urban air could have a detrimental effect.

Understanding that much of our particle pollution is created from chemical reactions between pollutants is vital to getting air pollution under control. When deciding which pollutants to target, we need to understand their toxicity or direct harmful effects and also the chemical reactions that they cause in the atmosphere. Yet we are faced with huge challenges. The absence of a direct link between the pollution that we produce and the secondary particles that we breathe makes them very difficult to control. The way in which air pollution travels between countries and the near invisibility of modern particle pollution makes it hard to convince city and national politicians alike of the need for action. It makes it easy for industry and polluters to resist, but the benefits could be huge.

Chapter 7

A tale of six cities

By the 1970s London's smogs were becoming a distant memory. The last great smog had occurred in 1962 and it looked as if the UK's urban air pollution problems had been solved. The world-leading Medical Research Council Air Pollution Unit at London's St. Bartholomew's Hospital was shut down. In the eyes of government, it had done its job. There was simply no further need for it.

At the same time as the UK was closing research capacity, in the United States, Doug Dockery and colleagues were setting up a new type of study to look at the health impacts of air pollution. When it was completed and published nearly twenty years later, Dockery's revolutionary findings changed our perspective of air pollution even more profoundly than London's 1952 smog.

Dockery did not start out as a doctor or health professional. His first degree was in physics and then gradually, via meteorology and environmental science, he moved to studying how the environment around us affects our health. Dockery and his team worked at the Harvard School of Public Health in Boston. Founded in 1913 to train public health professionals, the school's work has affected the lives of all of us around the globe. Starting with infectious diseases, their researchers invented the iron lung to keep polio victims alive, before pioneering a vaccine in the 1950s, which was recognized with a Nobel Prize. They also led the eradication of smallpox.* On a par with these achievements, the school can also list the air pollution work of Dockery and his team that began with the Six Cities study.

Starting in 1974, the Six Cities study involved 8,111 people who were selected at random from six chosen locations.[1] These were Watertown in Massachusetts, parts of St. Louis, the steel town of Steubenville in Ohio, and the less polluted towns of Portage in Wisconsin, Harriman in Tennessee, and Topeka in Kansas. Participants filled in questionnaires on their weight, height, smoking habits, occupation and medical history. Each had their breathing tested. Then, every year, Dockery's team sent a postcard to each person to find out if they were dead yet. If there was no reply, they sent an investigator to talk to family, friends and neighbors to find out what had happened. This went on for sixteen years. During this time 1,930 participants died. There was nothing extraordinary in this, but it was who had died, and importantly where they had lived, that mattered. The residents of Steubenville and St. Louis were dying faster than those in Topeka and Portage. Allowances were made for the number of smokers, body mass index and other factors that affect people's health, but there were still differences between towns that could not be explained. When these differences were plotted against the particle pollution in each city, however, an extraordinary pattern emerged.

* You can read more about the first hundred years of the Harvard School of Public Health at https://harvardmagazine.com/2013/10/100-years-of-hsph.

Cast your mind back to the laboratory experiments that you did at school and the results that you plotted on graph paper. There would always be a few wobbly points that did not quite fit. You would expect a sixteen-year experiment to contain many variables and anomalies, but the team found a near straight-line relationship that has gone on to become one of the most famous plots in air pollution science. It is replotted in Figure 2, using the original data from Dockery's team.[2]

Figure 2 Mortality rates and fine particle concentrations from the Six Cities study

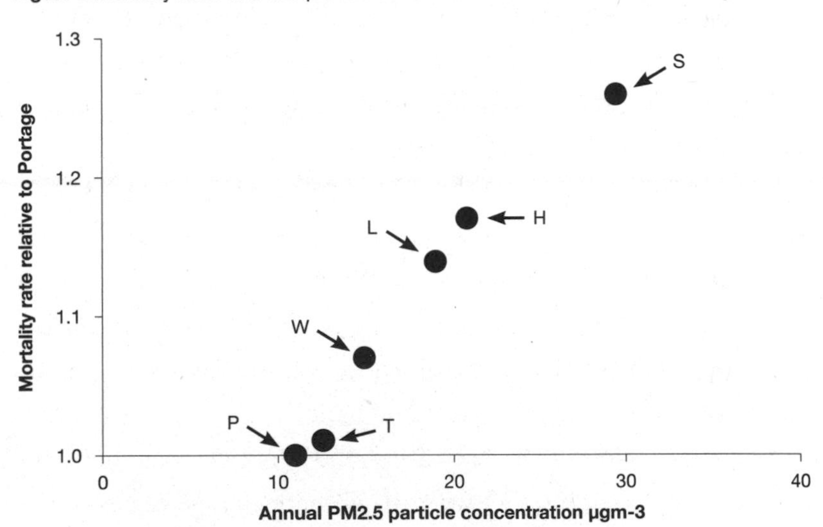

P = Portage, T = Topeka, W = Watertown, L = St Louis, H = Harriman, S = Steubenville

Due to the extra air pollution that they were breathing from the traffic and industry around them, people in Steubenville were dying nearly 30 percent faster than those in Portage. And people were not just dying from lung problems but, like those in London's 1952 smog, from heart problems too. This was not strictly speaking the first study to show that everyday exposure to modern air pollution was harming our health, but other studies had been less clear in their findings.

The publication of the Six Cities study in 1993 was followed swiftly by another that looked at death rates in a large group of US citizens who were being tracked to check the development of cancer.[3] They too were dying early because of the particle pollution that they were breathing. Suddenly we realized that the health impact of modern air pollution was bigger than anyone had imagined. Even looking at the cleanest of the six cities, the air pollution effect was still clear. Our air had to be made cleaner; cleaner even than the air in Portage. New effort was needed, not just in traditionally polluted places like the steel town of Steubenville, but everywhere.

New pollution laws would be required to protect people's health and set standards for the air in our cities. It was clear that industry and vehicle manufacturers, to name just two sectors, were going to have to do much more to clean up the pollution that they caused. But, as Arie Jan Haagen-Smit had found over forty years before, the vested interests would fight tooth and nail.

The publication of the Six Cities study in 1993 was met with controversy and debate. As with the climate change issues of the twenty-first century and the European acid rain problems of the 1970s, doubt was sown by those who did not like the results. How could Dockery be sure that it was the particle pollution that was causing people to die early? The people were breathing a whole range of air pollutants, not just one.* Dockery's team could not measure

* In fairness, other pollutants had been measured in the Six Cities study, but their links to the death rates were not nearly as clear.

every possible pollutant, so how about pollutants that had not been measured? Could one of them have been the cause? The particles in each city came from very different sources and had different chemical compositions, so how could they all produce the same effect? Association is not the same as cause; just because people in the more polluted cities died sooner, it did not mean that air pollution was actually causing their deaths.* Could the difference in death rates be due to some other differences between the cities; maybe it was differences in weather or the number of people who smoked?[4] Had the researchers simply made a mistake? How could tiny amounts of particles in the air cause people problems anyway? Was it really the mass of these particles in the air, or their number? Unless government could be completely sure, it should not unfairly damage company profits.[5] Less prominent in these debates were the voices pointing out that something was killing people in the hundreds in these cities and action was needed right away. Delay would mean more people dying prematurely.

One solution would be to carry out the Six Cities study all over again, to prove it a second time. But that would take at least sixteen years, plus time for analysis, during which time industry could carry on much as before. The US Congress got involved and it was decided that a separate and independent group of researchers would go through the data with a fine-tooth comb and repeat the original work.[6] Ultimately, in 2000, the original findings were confirmed. The everyday particle pollution that we breathe was, and still is, shortening our lives.

* This is a common problem for epidemiology. Just because things change together it does not mean that one causes the other. The researchers must weigh up the plausibility with other evidence.

The Six Cities study precipitated another evolutionary step in air management. We went from the 1950s approach of managing sources to new air pollution laws that set standards for the air itself. Standards and limits were created in the United States and Europe, and guidelines were set by the World Health Organization.

The impact of the Six Cities study did not stop there.[7] Between 1990 and 1998, a further 1,394 of the original 8,111 people died. Remembering the controversy from the original study, the researchers returned to all of the survivors to interview them, weigh them and find out their current health and smoking habits. Once again, the same straight-line relationship between survival and particle pollution was found, but this time there was better news too. During the 1990s, action had begun to improve air pollution. Over the whole twenty-six-year study period, particle pollution had fallen, and the greatest improvements were seen in the most polluted cities. In Steubenville and St. Louis, particle pollution in 1998 was at less than two-thirds of the level measured twenty years before. At the other end of the scale, Portage and Topeka showed little or no improvement. Where the air pollution improved, the survival rate improved too.* Cleaning the air worked and at least some of the harmful health effects were reversible.

The Six Cities study looked at adults only, but what about the impact of air pollution on children? As Dockery and team were completing the Six Cities study at the Harvard School of Public Health, over in southern California, Jim Gauderman and colleagues

* Researchers found a similar result when they checked on the Six Cities survivors again in 2009. See Lepeule, J., Laden, F., Dockery, D., and Schwartz, J. (2012), "Chronic exposure to fine particles and mortality: an extended follow-up of the Harvard Six Cities study from 1974 to 2009." *Environmental Health Perspectives*, Vol. 120 (1), p. 965.

began studying the effects of air pollution on children. Gauderman started his study in 1993 as a fresh graduate with a PhD and he still continues his work today.

Gauderman's study was different from the Six Cities. With their lives ahead of them, the team could not wait until the children grew old and died. Instead they looked at the way in which the children's lungs grew. In their first study, they studied over 3,000 children from schools in places with different levels of air pollution. Their lungs were tested twice over a four-year period. Their height and weight were measured and their medical histories were checked. The team looked at each child's home too, checking for second-hand tobacco smoke, fumes from gas cooking, damp and even cockroaches. They also measured the air pollution that the children breathed in each of the towns and neighborhoods. As in the Six Cities study, they found that it was not just smog that we needed to worry about; the air pollution that children breathed was affecting the way that their lungs grew. The lungs of the children in the most polluted places were growing more slowly than those in the least polluted areas; a difference of between 3 and 5 percent in the rate of lung growth. Astonishingly, the impact of air pollution was nearly five times greater than the impact from second-hand smoke at home.[8]

Over a period of twenty years, Gauderman went on to study more groups of children.[9] As with the Six Cities study, air pollution improved over time and there was some good news to be found. The children in the later studies who breathed less pollution developed slightly larger lungs. Again, this demonstrated that reducing air pollution yielded positive results.

Knowledge that air pollution affects the very young was not new. Children were among the groups that suffered most in the 1952 London smog,[10] but finding that air pollution caused permanent damage in children gave a whole new dimension to the health consequences of air pollution. Damage to children's development, stunting their lung growth, meant that we could be storing up a legacy of harm that might not manifest itself for decades—until the children of today reached old age. These lifelong impacts of air pollution were highlighted in a 2016 report from the Royal College of Physicians.[11] Not only did they report damage to children's lungs but also possible health effects from air pollution that could begin in the womb.

The lifelong effects of pollution exposure are very hard to investigate. If we started a study now it would take years before we knew if today's air pollution was damaging our health, and by the time we had our results, it would be too late to do anything about it. Another way to proceed is to look backward at people's lives and the air pollution that they breathed in the past. Starting in 2008, Anna Hansell and her team at Imperial College London spent around a decade mining the old data sets of air pollution measurements in the UK. Hansell worked in the Small Area Health Statistics Unit (SAHSU), hidden away in a corner block at St. Mary's Hospital, London, just a few meters from the location where Alexander Fleming had discovered penicillin.

The unit did not start with air pollution research. It owes its existence to a television documentary in 1982 when journalists from Yorkshire Television found high rates of leukemia in children and young people who lived close to the Sellafield nuclear fuel reprocessing plant in Cumbria. The Sellafield plant includes Windscale, which was the location of an infamous radioactive fire in 1957. The government inquiry that followed confirmed the cluster of leukemia cases but could not confirm that the nuclear plant was the cause.[12] Conscious that small pockets of illness might be going unnoticed in other places, the government set up SAHSU as a permanent unit to look at health statistics around industrial facilities to provide early warnings. In the last twenty-five years they have investigated the health impacts on the people living near power lines, landfill waste sites, mobile phone masts and waste incinerators, along with air pollution and aircraft noise.

To look at the long-term impacts of air pollution, Hansell's team randomly selected 1 percent of the population from the 1971 census—a total of 370,000 people, including children and the elderly. They then searched for the same people in each successive census to see if they were still alive. It would have been impossible for Hansell's team to visit the families of each person who had died in the way that Dockery's team had done. Instead they accessed electronic copies of death certificates. They also estimated the air pollution that each person had been exposed to. After a lot of computer processing, Hansell's team found that the main risks came from the pollution breathed in during the last decade of their life; but, astonishingly, the air pollution breathed before this also affected people's survival, right back to the air pollution that they had been exposed to nearly forty years before.[13]

As with many diseases that involve a time lag between exposure and death, the lack of an immediate effect does not mean that we should be lulled into a false sense of security. We need to redouble the action that we take now for the sake of our own later lives, and most importantly for the sake of our children and those yet to be born.

Taking action is not so easy. Which source of particle pollution should we tackle first? If we could figure out which part of the particle mix was the most harmful we could tackle it more efficiently and more quickly. Here the health studies do not help us very much. In the years since the Six Cities work, numerous studies have pointed toward one source or another as being the most harmful but there is little consistency in their findings. Dockery's Six Cities study pointed the finger toward sulfate particles that form in the air from the burning of coal and oil as the main cause of the early deaths. Hansell's UK census study would suggest soot particles or sulfur dioxide. Another theory suggests that the critical issue is the chemical reactions that particles cause on the surface of our lungs, overwhelming our natural defenses.[14] This would lead us to look at metal particles from vehicle brakes. I could go on, and the list would be long.

One alternative is to change our thinking and stop looking for the single most harmful pollutant. We never breathe one pollutant at a time; we breathe a mixture. So perhaps we should be investigating what mixtures of air pollution produce the greatest health impacts, rather than which single pollutant? It could be traffic pollution, wood-burning or the mixture of pollutants from coal-burning. To use a metaphor, we should be trying to see the wood from the trees. This was the topic of a study that I undertook with my PhD student Monica Pirani. It was a monumental battle with statistical modeling. We found that the greatest risk—having allowed for temperature and other factors—for people dying of breathing problems in early twenty-first-century London was on the (mostly) springtime days when particle pollution forms over wide areas of northwest Europe.[15] These are the same pollution mixtures that Owens would have breathed on his Norfolk holiday and that caused the Scandinavian forest dieback and fish deaths of the 1970s and 1980s. This finding would point to taking action on traffic, coal-burning industry and agriculture all together. However, these types of study are in their infancy.

As researchers, air pollution scientists tend to focus on the remaining research questions rather than on what is already known. This should not be interpreted as doubt or reason for delay or inaction. There have been many thousands of studies around the world. The evidence that particle pollution shortens lives is unequivocal and air pollution exposure is the largest environmental risk factor for early death worldwide.[16]

It is hard to overstate the impact of the Six Cities study on global health. Nearly twenty-five years after its publication the results still offer the best estimate for how much our lives are shortened by the particle pollution that we breathe. In 2017, the biggest air pollution study to date looked at over sixty million Americans enrolled in Medicare and further confirmed the results of the Six Cities study.[17] Thanks to the work of the Six Cities team* we can now estimate that particle pollution contributed to over four million deaths—not all of these were old people nearing the ends of their lives—around the world in 2015, representing an astonishing 7.6 percent of global deaths.[18] However, the legacy of the Six Cities researchers lies not in these numbers but in the lives saved by actions that have been taken so far to clean our air, and actions that will be taken in the future.

* Douglas W. Dockery, C. Arden Pope, Xiping Xu, John D. Spengler, James H. Ware, Martha E. Fay, Benjamin G. Ferris Jr. and Frank E. Speizer.

Part 3

The battle grounds of today: Modern problems in a modern world

Chapter 8

A global tour in a polluted world

In Chapter 1 we met some of the early explorers of our air, including John Aitken and Robert Angus Smith, who both traveled around taking air samples with their own handmade equipment. Aitken made many measurements across Scotland, journeying from his home in Falkirk. Smith traveled out from his laboratory in Manchester. In keeping with the fashion for European tours at the time, both Aitken and Smith traveled around the continent, mainly in France, Switzerland and Italy. Their explorations were like those of Victorian plant hunters: climbing mountains and, in Smith's case, exploring mines and underground train tunnels, capturing specimens as they went. They would have traveled by boat and train, taking advantage of the new opportunities for travel in the age of steam.

Perhaps the best-known explorers of the late nineteenth century, however, are two people who never actually existed. In October 1872, the same year Smith published his book *Air and Rain*, Jules

Verne's fictional characters Phileas Fogg and Jean Passepartout set out from London to circumnavigate the globe. A theoretical grand tour around the world today, following the path set out in *Around the World in Eighty Days*, would reveal the global magnitude of the air pollution problem, the contrasts from continent to continent and some of the environmental inequalities at the heart of our air pollution problems.

Let's start where Fogg began, in London. The London area is much bigger now than it was in Fogg's time, mainly due to the suburbs built in the 1920s and 1930s. Including outlying towns it has a population of more than 10 million, making it one of Europe's two megacities. Today London is very different from the coal-dominated city of the past. Its air and buildings are no longer blackened by smoke; instead its air pollution problems are dominated by traffic, especially from diesel vehicles. Without an extensive streetcar system, and with an underground railway system that mainly runs north of the Thames, the city is heavily reliant on buses. Around half of London cars are powered by diesel. This is a very European peculiarity; diesel cars are almost absent from the rest of the world. The popularity of the diesel car, van, truck and bus in Europe means that nearly three-quarters of the world's diesel vehicles are driven on Europe's roads.[1] Despite ever-tightening exhaust controls, emissions from diesel vehicles in the real world are very different from the way in which they perform in laboratory tests, meaning that London wrestles with air pollution problems, including nitrogen dioxide and particles. London does not meet World Health Organization guidelines for air pollution. In 2017, some roads still exceeded European pollution laws by a factor of more than two, and for many years annual air pollution limits have been breached within the first week of January. Air pollution is hotly debated in the capital's media, especially in the *Evening Standard* newspaper, and from the Londoner's perspective a lot more needs to be done.

Step outside the capital and a different view emerges. From an international perspective, London is seen as a world-leading air pollution innovator. Novel schemes include the congestion charge, which levies a toll on drivers in a small central zone and uses the money for public transportation. London also has the world's largest low emission zone, which bans the most polluting trucks and buses.

Britain's island status influences its worldview. It also determines the country's view of air pollution. However, Europe is just a few hours away on the drifting air. It is better to think of London as part of the westerly side of one of the most populated parts of Europe; a packed region that straddles the English Channel and North Sea. This includes southern England, Belgium, the Netherlands and the industrialized regions of northern France and the German Ruhr. All these places share the same polluted air.

London is only 200 miles from Jules Verne's home in Paris, Europe's other megacity. With particle pollution staying in the air for a week or more, the circulation of air is a critical factor in Parisian air pollution. Comparatively, the UK is lucky. The quality of its air is helped by a dominant westerly wind from the Atlantic that mainly carries air pollution from the UK eastward, across Europe;[2]* this explains why the UK was Europe's largest net exporter of air pollution in the acid rain–dominated 1970s and 1980s (see Chapter

* A great perspective on Europe's air pollution can be obtained from the European Environment Agency's annual reports.

6). If we were to follow the prevailing wind eastward, we would find that, unsurprisingly, air pollution tends to worsen. This is due both to the accumulation of pollutants and, as we reach Eastern Europe, the greater use of coal in industry and home heating. Poland is an air pollution hotspot thanks to its coal use.

Further to the north, air pollution across Scandinavia is generally less than in many parts of Europe. There are fewer people to emit air pollution, and the weaker sunlight here means less energy is available to drive chemical reactions in the air. However, Scandinavia suffers from a uniquely northern European problem caused by winter sanding of roads and by studded winter tires. In winter everyone changes their car tires. Many of these have tiny metal football-type studs which grip on the ice and compacted snow of a Scandinavian winter, but also wear the roads to dust. Walk along the streets of Scandinavian or Icelandic towns and cities in spring and you will see the gutters filled with road dust, clouds of it being stirred up behind trucks and buses. Street sweeping is of limited use, as each passing vehicle grinds new dust out of the road surface. Chemicals can help to suppress the dust, but the longer-term solution looks to be a swap of metal-studded tires for rubber tires with improved winter tread. With plentiful access to forests, winter across Scandinavia also means wood heating and therefore smoke rich in soot particles and organic chemicals.

Fogg and Passepartout headed to the Mediterranean by train. Here the sunlight has more energy to drive the chemical reactions that create some types of air pollution. The dryer summer climate means windblown dust worsens the particle pollution problems. The nearby Sahara increases Mediterranean woes, as desert dust often adds to the pollution burden. Unlike the UK and the Netherlands, in Western Europe, some countries have big tax incentives on diesel fuel and the vast majority of their cars now run on it. Small motorcycles that prevail in Mediterranean countries lack the modern exhaust controls found on larger vehicles, adding to urban air problems.

Traveling by train, Fogg and Passepartout skirted the Alps. Here, particles from wood-burning and traffic buildup each winter in the cold, deep valleys. This leads to pollution problems in many small Swiss towns, in Grenoble in France and along the Rhône.

The train then took them along the Po Valley in northern Italy. As we have seen, this is one of Europe's most polluted regions. Here, population and industrial density, gentle winds and strong sun lead to the perfect conditions for air pollution to linger and stew. This creates problems with ozone forming close to the ground, particle pollution and nitrogen dioxide.

From Brindisi in Italy, Fogg and Passepartout sailed to Suez and then down the Red Sea, bound for Bombay. Across the Middle East, the dry climate and dust blown by the wind adds to problems from the petrochemical industry, similar to those that Arie Jan Haagen-Smit found in Los Angeles. This means that particle pollution and ozone prevail across the region. Dust storms are seen as part of the natural environment here, but they are not harmless. They are thought to cause respiratory problems, cardiovascular complaints, meningococcal meningitis, conjunctivitis and skin irritation, on top of the more obvious deaths and injuries from road accidents caused by reduced visibility.[3] The dust is often made up of mineral-rich particles that can be harmful if they enter our lungs. Those of us who live in the wetter parts of the world, with few dust storms, tend to think of desert dust as being like the sand that we might use to make concrete. It is more accurate to think of dust storms as windblown soil. It is not sterile like construction sand, it contains lots of harmful biological and plant material. Some of these dust storms, moreover, are not entirely natural, being caused by the way land is managed for agriculture and by water extraction from rivers and lakes, leaving the soil vulnerable to erosion by the wind. And desert dust can travel a long way. In May 2007, a dust cloud from the Taklamakan Desert in China made a complete circuit of the globe in just thirteen days.[4]

To the east of the Red Sea and the Persian Gulf is Iran and its capital, Tehran, a global air pollution hotspot. But it is further east, across India and East Asia, where we find the world's worst air pollution. India and China are the two most populous countries in the world, and the region, which also includes Pakistan, Bangladesh and Indonesia, is home to around almost half of humanity. Here cities experience the traffic pollution of the developed world, the poorly regulated industry typical of the developing world and the open burning of wood and rubbish for cooking that we find in the world's poorest areas—often all on the same street. The people in India and Bangladesh have the highest and fastest rising exposures to particle pollution.[5] The huge amounts of coal burnt across China mean that Beijing has replaced London as the world's archetypal smoggy city. To add to the country's woes, much of China also suffers from desert dust from inland dry regions.

After crossing India, Fogg and Passepartout sailed through Indonesia. Agricultural burning or—perhaps more accurately—

forest and peatland burning here can cause pollution problems far afield across Singapore and Kuala Lumpur. These are not natural fires but are part of land management and forest clearance. The resultant air pollution across Indonesia and East Asia is estimated to cause up to 300,000 deaths per year during peak El Niño years when the fires are at their worst.[6]

Hong Kong was part of the British Empire in Jules Verne's day. It is now a special administrative region of China and one of the world's largest ports. At this point on our global grand tour it is worth mentioning shipping as an air pollution source. The boats on Fogg and Passepartout's adventure were coal-powered steamers. Today shipping is powered by heavy fuel oil, rich in sulfur. Sample the air anywhere in Europe and you will find vanadium, a metal found in shipping fuel that finds its way into the exhaust from ship funnels. The impacts of shipping pollution are much worse in coastal areas and port cities. In Hong Kong, shipping pollution blowing across the city has been associated with increases in emergency hospital admissions for heart attacks and cardiac problems.[7]

Next on Fogg and Passepartout's route was Japan, before they crossed the Pacific to California and journeyed overland to New York. Travel around the developed world and you will find that cities are full of traffic. You would therefore think that they have similar pollution problems, but there are important differences. Gasoline-powered cars dominate in Japan and the United States, compared with Europe, where diesel is king. A few years ago, I played host to some Japanese scientists who were visiting the UK. We each presented air pollution data from our cities and I was shocked by the Tokyo measurements. Particle pollution and nitrogen dioxide next to their gasoline-dominated roads was a fraction of what I was measuring in the diesel-filled streets of London. City heating also differs. The availability of natural gas in Europe means that it is the fuel of choice, whereas the skyscrapers of New York are heated by oil, leading to heavy metal and sulfur particle pollution.

Once synonymous with summer smogs, the air of Los Angeles and San Francisco is better today than it was in Haagen-Smit's time. This is thanks to strict controls on vehicles and industry, and the work of the California Air Resources Board, but the battle is not won even here. California still experiences the worst ozone pollution in the United States, and the worst particle pollution too. Particle pollution can be different even within one country. In the United States it is dominated by traffic on the west coast and by industrial coal- and oil-burning on the east. The colder parts of the United States have to deal with additional problems from winter wood-burning, particularly around Seattle and Montana.

The shortest route back across the Atlantic takes ships and aircraft to the north, following a great circle. Although Fogg and Passepartout did not take a northerly detour to Iceland, it is worth pausing here on our grand tour. Television and radio news readers have had to learn to pronounce the names of Icelandic volcanoes over the last few years. In 2010, dust from the Eyjafjallajökull eruption confined aircraft to the ground across much of northwest Europe, and Grímsvötn did much the same in 2011. Dust from the Grímsvötn volcano caused particle pollution across the UK and Europe, but it is not just the dust from volcanoes that should concern us. The eruption of the Bárðarbunga volcano in 2014 to 2015 spread sulfur dioxide gases across the UK and Ireland that could be smelt in Norway.* This was, though, a minor incident compared to the Laki eruption of 1783 to 1784, when huge amounts of sulfur gases spread across Europe and formed tiny sulfate particles. Leaves withered and fell from the trees. Deaths across England increased by between 10 and 20 percent and there are records of breathing problems and increased deaths across the Netherlands, France, Italy and Sweden. A repeat of this eruption today would add an estimated 142,000 to Europe's annual air pollution death toll.[8]

A southerly detour would take us past some of the Atlantic islands. These include both Tenerife and Cape Verde, home to monitoring sites that are part of the Global Atmosphere Watch. The organization is comprised of over thirty stations, all located in remote places; the most famous is Mauna Loa on Hawaii, which has been tracking the global rise in carbon dioxide for decades. Together, this remote network quietly watches the changes that human activities are causing to the composition of our atmosphere. These changes include increases in carbon dioxide that are driving climate change, but the network has also found that methane leaked from fracking operations in the United States is spreading around the world. Monitoring sites on Tenerife, Cape Verde and the Azores, as well as those on top of the European Alps, have all detected leaking gas from fracking in the United States.[9]

After their transatlantic crossing, Fogg and Passepartout made landfall at Queenstown, now called Cobh (pronounced "cove"), near Cork in Ireland. This was the *Titanic*'s last port of call and the embarkation point for much of Irish emigration into the United States.

* The spread of sulfur dioxide across UK and Ireland is briefly discussed in https://www.theguardian.com/environment/2014/sep/28/pollution-iceland-ireland-sulphur-dioxide.

Ireland has some unique air pollution problems. Its westerly position means that it normally receives a clean air flow from the Atlantic, but its distance from the North Sea and European mainland meant that natural gas grids developed late here. Pipelines only reached Ireland in 1990, more than twenty years after homes on the UK mainland started to enjoy gas central heating. Although it is one of the very few countries to meet European legal limits for air pollution, the lack of widespread natural gas leaves the winter air of Ireland's smaller towns dominated by particle air pollution from home heating with coal and peat, creating a unique Irish pollution challenge.

Fogg and Passepartout finished their eighty-day journey by traveling from Ireland to Liverpool, finally arriving back in London by train. Their route around the world stayed almost entirely in the northern hemisphere, where they could take advantage of coastal and overland trade routes. With greater expanses of sea, and less land, a tour of the southern hemisphere would have been difficult for Fogg and Passepartout, even in the age of coal and steam. Global weather patterns keep northern and southern air apart and there is only limited mixing between the two. Describing the air of the southern hemisphere remains challenging today since pollution measurement systems are far less developed there.

There are relatively few measurement sites in the Middle East and large parts of Asia and Africa. In 2015, the whole of Africa had just fifteen monitoring sites;[10] the single city of Paris had more than three times as many. Indoor burning of solid fuel for cooking adds hugely to the air pollution burden across many areas of the African continent. Indoor burning of dung, wood and coal adds a further 2.85 million deaths to the global air pollution toll, making a massive impact on child mortality from diseases such as childhood pneumonia. There are problems in the more industrialized areas of Africa too. One example is the Niger Delta, where flaring from oil and gas production causes air pollution that drifts several hundred miles inland.[11]

Australia and New Zealand enjoy healthier air than much of the globe. The Australian government frequently points out that its air is relatively clean by world standards, but that is not to say that it is as clean as it could or should be. According to the International Council on Clean Transportation, Australia lags behind many developed countries in removing polluting sulfur contaminants from road fuel.[12] In Sydney, particle pollution is estimated to cause around 430 early deaths in the city each year, as well as a further 160 from ozone, along with over 1,000 hospitalizations.[13] Forest fires add to the city's pollution problems. Progress in cleaning Sydney's air has stalled in recent years and it may come as a surprise to learn that around half of the particle pollution comes from home wood-burning. This problem is much more severe in the more southerly, and colder, Tasmania.

New Zealand is of a similar size to the UK but has less than 10 percent of the population. With around 1,200 miles separating New Zealand from any large land mass we would expect low air pollution. Most images of New Zealand show a pristine environment of great beauty and pure, clean air. The reality is different. Airborne particle pollution in many towns exceeds WHO guidelines. The country's air pollution problems are very much homegrown. This is not due to the diesel cars that confound efforts to manage air pollution in Europe, or to the density of cities and industry that contribute to problems in East Asia, Europe and parts of North America. It is mainly due to home heating.

With the country's limited coal and natural gas, and with expensive electricity, many New Zealanders, especially those in the South Island, rely on wood-burning to heat their homes. National standards for particle pollution are breached in Christchurch and even in small towns right across the South Island.[14] Poorly insulated homes and fuel poverty contribute to high winter death and childhood asthma rates.

The Clean Air Institute highlights that 100 million Latin Americans are living with air pollution above WHO guidelines.[15] Early deaths and the cost of looking after people who are made ill by air pollution results in a burden of between two and six billion dollars per year on the economies of the continent. In a region of contrasts, some countries have no frameworks for controlling air pollution while some cities, including Mexico City, Bogotá, São Paulo and Santiago, have shown great improvements. As in much of the world, particle pollution, ozone, poor controls on industry, fuel and traffic are the main problems. The chemical composition of the air in São Paulo, South America's largest megacity, stands out due to Brazil's response to the oil crisis in the mid-1970s.[16] The leap in oil prices coincided with a dip in the price of sugar cane, an abundant crop in Brazil. This perfect storm created an ideal opportunity for the creation of a national ethanol program. Cars went from running on gasoline containing only 10 percent ethanol to cars running on almost pure ethanol. The fall in oil prices in the 1980s and the discovery of offshore oil led to a reversal. These fluctuations mean that Brazilian cars are typically "flex-fuel," and their owners are able to swap between gasoline and ethanol fuels as prices change. Compared with the rest of the world, this use of different fuels means different air pollution, and increased ethanol consumption has been linked to worse ozone pollution. Sugar cane contributes to air pollution in other ways. Following the harvest between May and October, farmers burn the crop waste and large areas of Brazil can be affected by smoke.

So, a tour of the modern world reveals that each city, country and region has its own challenges. Let's look at some of these in more detail, starting in the most famously polluted place of the early twenty-first century, Beijing.

The Chinese capital gained its recent reputation in the months before the city hosted the 2008 Olympic Games. Concerns had been expressed about how the athletes would be affected by air pollution, but there was almost no hard data available. This changed abruptly in July 2008 when a simple Twitter message appeared from the roof of the US embassy in Beijing. As a service to US citizens, the embassy had installed equipment to measure particle pollution. Rather than using the data to produce reports that no one would read, the embassy staff connected the equipment to Twitter. Suddenly Beijing's air pollution was instantly visible to the world. Each hour a tweet from @BeijingAir would appear, much like a message from a caged canary in a coal mine. This equipment, and most importantly its Twitter feed, changed the course of air pollution management in China and around the world. It catapulted China's air pollution from government secret to front-page news. Instead of simply giving numbers, the US embassy translated measurements into messages that described the quality of the air as "good," "moderate," "unhealthy for sensitive groups," "unhealthy," "very unhealthy" or "hazardous," using health recommendations from the US Environmental Protection Agency (EPA). Instantly an objective judgment was being made on Beijing's air and broadcast to the world without any diplomatic niceties. The Beijing government tried to have the messages taken down, declaring them illegal, but the embassy stood its ground.[17]

The media attention reached a peak when, in 2010, the measurements from the US embassy went off the top of the EPA scale, going from "hazardous" into uncharted territory. The next tweet simply said, "crazy bad." Beijing began to replace London as the archetypal smoggy city and pressure on the Chinese government grew. Something had to be done.

In 2012, new Chinese laws were enacted and 138 monitoring stations in 74 cities began publishing air quality data, with trial runs in another 195.[18] The picture that they revealed was not a healthy one. Northern cities were more affected by sulfur dioxide from coal-burning than those in the south, but ozone affected the whole country. Particle pollution in the average Chinese city was almost six times above World Health Organization guidelines. Some sources of particle pollution were new and were the result of the country's rapid industrialization. A huge amount of coal was being burned with few controls in newly built power stations and factories. Older sources contributed too, including agricultural burning and home heating. Pollutants were mixing together to form secondary particles and ozone, enveloping whole regions of the country. So the solution to Beijing's air pollution did not lie wholly within the city. Pollution controls were also needed across neighboring regions.

In January 2013, Beijing was experiencing its worst smog since 2008. The US embassy reported a "crazy bad" warning again and it was no longer possible for the government to ignore what the people could see, smell and taste for themselves. Overnight there was a sudden and huge change in the way that air pollution was reported in the Chinese media.[19] Critical articles appeared with headlines like the *China Youth Daily*'s "Lack of Responsive Actions More Choking Than the Haze and Fog," and a new attitude unfolded.

The Chinese investment in air pollution measurement was astonishing. In 2012 there were no public measurements. By 2014 a network of over 1,300 monitoring sites was operating across 367 cities. The network was around ten times larger than that in the UK and was built in just two years. It revealed that Beijing was not the most polluted place in China, as media reports suggested. Particle pollution problems were most severe around the rapidly growing megacity clusters. Provinces such as Hebei and Tianjin topped the list, but the problems were widespread; 1.3 billion Chinese were breathing air that did not meet WHO guidelines. Air pollution was increasing the death rate by around 15 percent nationally (around 1.7 million early deaths per year), and was implicated in around 40 percent of stroke deaths.[20]

Sadly, China shows the price that is paid by pursuing economic growth at the expense of environmental impact, but we cannot blame industry alone. Many of China's air pollution problems predate its massive industrialization. The new measurement systems also revealed that, in air pollution terms, China is very much two countries in one, divided by the Huai River and the Qinling mountain range. This dividing line is not due to a massive geographic feature but results from a central government policy dating to the 1950s. The Huai River and the Qinling range mark the line where average winter temperatures are at 32°F. In the colder north, free or highly subsidized coal was provided to homeowners, and in many towns and cities, highly polluting coal-fired district heating systems were built. This did not happen in the warmer south. The policy has been described as having had "disastrous consequences for human health."[21] North of the Huai River and the Qinling Mountains, the extra air pollution from coal-burning shortens life on average by 3.1 years. People in the north tend to die earlier from heart and lung problems consistent with air pollution exposure. The consequences of China's air pollution are so great that the government had to act, and in less than a decade China went from concealing its air data to being on the cusp of leading global air pollution control.

The same lack of measurements that obscured the nature of China's air pollution problem manifests itself in many other areas of the world. Although legislation has driven the development of measurement networks in the European Union and across North America and Japan, the availability of air pollution data for much of the rest of the world is sparse, and for many areas it is almost nonexistent.

No method for measuring air pollution will be perfect, and none will have global coverage, so how can we make best use of the imperfect data that we have? This was the job of Gavin Shaddick from the University of Exeter. Shaddick's contribution to mapping the world's air pollution is very different from the early explorers of the Victorian era, who took air samples as they traveled. It is very different from the twenty-five years' worth of measurements I have made with my team in London. Shaddick did not travel the world, or indeed leave his desk. Shaddick is a statistician. Working for the World Health Organization, he took global satellite data, measurements from the ground, and pollution estimates from computer models to make an air pollution map of the world.[22]

Reading newspaper headlines about air pollution, you would expect the worst problems to be in Beijing and across Europe and North America. Shaddick's map showed a belt of particle pollution that stretched from West Africa, across the Sahara and the Middle East (where desert dust adds to the particle pollution), through northern India (especially the areas around the Ganges) and across China. Next, the global population was placed on the map and the impacts of air pollution on humanity could be estimated.

The numbers were astonishing. In 2016, 95 percent of humanity breathed air that did not meet WHO guidelines and the situation was getting worse, especially since the turn of the century. The most extreme particle pollution was experienced by the populations of China, India, Pakistan and Bangladesh. The impacts of particle pollution increased from around 3.5 million early deaths in 1990 to 4.1 million worldwide in 2016; over half of these were in China and India.[23] The biggest increases were not in China but across India and Bangladesh. Globally, breathing particle pollution was the sixth greatest risk factor for early death, just behind high blood pressure, smoking, high blood sugar, being overweight and high cholesterol. In better news, some situations had improved. Annual early deaths in Europe decreased from around 330,000 to 260,000, but this was still more than three times the numbers in the United States. Nigeria improved from 77,000 to 51,000 early deaths per year over the 1990 to 2015 period.

Ozone, the pollutant at the focus of Haagen-Smit's work, took its toll too, contributing to 254,000 early deaths in 2015, which made it the thirty-third highest risk factor for early death. India was the country with the greatest deterioration since 1990, accounting for 67 percent of the global increase in ozone deaths.[24] Ozone concentrations in the populated parts of the world increased by around 7 percent between 1990 and 2015, but the changes have not been the same everywhere. Ozone across North America decreased, and only small rises were seen in Europe. The largest increases were in the massively populous countries of Southeast Asia and in Brazil.

As Haagen-Smit noticed, ozone affects plants and crops too. Globally, ozone leads to losses of between 7 and 12 percent for wheat, 6 to 16 percent for soybean and around 4 percent for rice and maize. Ozone reduces European crop yields by around 2 percent, but the impact in India and the surrounding countries is far more significant. This area is home to almost one-third of the world's undernourished people and ozone is reducing their crop yields by up to 28 percent. The increase in ozone crop damage could be one reason why India's efforts to produce more food are resulting in a sluggish growth in crop productivity. There is serious damage to crops that are important to the local diet. For instance, the peas and mung beans that are important sources of protein in largely vegetarian India may be experiencing yield losses of 20 to 30 percent.

It is not just food crops that are affected. Globally, ozone is likely to be reducing tree growth, damaging the timber industry and reducing the rate at which trees absorb carbon dioxide, an important air pollution and climate interaction. Not only does the ozone damage burden fall unevenly around the world, but ozone-forming pollutants emitted in one region can often have impacts in another. One example is the reduction of European crop yields due to pollution from North America.[25]

Haagen-Smit's investigations of ozone told us that it mainly came from traffic exhaust and oil refining and happened during hot, sunny weather. As we understand this pollutant more we can see that the global background is rising, much the same as it is for carbon dioxide. The ozone pollution in Paris is around twice as high today as that measured over 100 years ago. Across the temperate regions of the northern hemisphere, ozone now increases every spring thanks to pollutants that have built up during the winter months, ready to react in the stronger spring sunshine. The southern hemisphere is less industrialized, but pollutants produced from land and forest fires are important contributors to the ozone seasons across the tropics. This means that ozone has become a global issue in the same way as climate change.[26]

In 2008, the UK Royal Society called for an international agreement to manage the pollutants that cause ozone across the world.[27] So far no one has listened. In contrast to the consensus on controlling ozone in the stratosphere, there has been no agreement to curtail the ozone that is damaging our health and crops at ground level. The notable exception is the Gothenburg Protocol, established during the Cold War to control acid rain. This also covers some ozone-forming pollutants, but it is restricted to Europe and North America. Outside Europe, the United States and Japan, there are few restrictions on industrial pollution. Methane, one of the main gases that forms ozone, is only lightly controlled despite its role in climate change. Even in the developed world, methane from farming and old coal mines falls outside regulation, as does most wood- and forest-burning. The situation is getting worse, not better, as uncontrolled sources of ozone-forming pollutants increase, including those from the coatings, printing inks, adhesives, cleaning agents and personal care products that we use in our homes.[28]

A worrying development emerged in 2009 when a new type of ozone smog appeared across the Uinta Basin in the US state of Utah.[29] This is a large flat area bounded by mountains to the north and east, and its winters can be cold. During one notable event between twenty and thirty centimeters of snow covered the ground, but ozone rose to the concentrations normally only seen in hot summers. It could not be more unlike the conditions that led to the Los Angeles smogs. Something similar had been seen five years before in neighboring Wyoming, as a result of which the state breached US ozone standards. With this in mind, researchers in Utah set about finding the source. They started with the air pollution sources that they knew about, but it simply did not add up.

As in Wyoming, the shales in the Uinta Basin in Utah had recently been developed for fracking. This involves injecting liquids underground to fracture the rock in order to extract oil and gas. Huge controversy, including the arrest of the UK Green Party leader Caroline Lucas, surrounded the drilling of just two pilot wells in the UK—one in the northeast of England, near the seaside town of Blackpool, and another in the southeastern county of Sussex. According to the *Washington Post*,* 137,000 new wells were drilled in the United States in just six years from 2010 to 2016. It is impossible

* An article with great infographics, including pictures of flares as seen from space, can be found at https://www.washingtonpost.com/graphics/national/united-states-of-oil/.

to estimate how much gas leaked by looking individually at each well and the miles of pipework, pumps and machinery. Instead researchers have flown over the oil and gas fields with instrumented aircraft, measuring what was coming from the ground below them. One flight over part of the Uinta Basin revealed the answer to the winter ozone. A lot more methane was leaking from the wells than had been thought, an extra 40 percent. In winter, this was being trapped in the cold air layers close to the ground, forming ozone as the low-angle sun was reflected off the snow.

More flights over US shale gas fields revealed large methane sources, but these areas also have cattle farms that produce methane. The shale gas plants can simply blame the farmers. The two sources therefore need to be separated in the data. Usefully for this type of experiment, shale gas also contains ethane, which does not come from natural sources such as farming. Looking at ethane and methane together showed that shale gas and oil extraction were overwhelmingly the dominant source, not farmers. Some active drilling areas were notable "super-emitters," suggesting that this phase of the shale gas production is the worst of all.[30]

Ethane can remain in the air for months, making it useful as a means of tracing leakages from natural gas as air travels around the world. The Global Atmosphere Watch has been measuring the composition of our air for over thirty years. One of its sites is located on top of Jungfraujoch in the Alps. Overall the news from here had been good. Better controls on the European gas industry had been leading to slow reductions in ethane since the 1980s. But suddenly in 2009, around the start of large-scale fracking in the United States, the ethane trend reversed.[31] Ethane started to increase, and not by a small amount. It was increasing at 5 percent per year. This points to a big increase in the global methane leakage from natural gas use.

Looking at measurements from locations in remote places around the world, notable differences can be seen.[32] Ethane is not increasing everywhere. At Lauder, on New Zealand's South Island, the trend has continued gently downward in common with most of the southern hemisphere. East of the United States, at monitoring sites on Atlantic islands and across Western Europe, it is a different story. All these places have seen increases. Generally, air flows in an easterly direction around the world, so it looked as if the new sources lay in the United States. This was confirmed by looking at propane, which is also found in oil and gas but has a shorter life in the atmosphere. Again, it was the Global Atmosphere Watch sites on the east of the United States and in the Atlantic* that detected increases, but the gradient across the United States was striking. There was a decrease in propane on the western side of the United States but on the eastern side it had increased sharply. There can be little doubt that the huge expansion of natural gas and oil production across the United States led to these increases. US methane leakage may be around twice the official estimates and it is having a global impact. There is clearly a need for better controls.

By 2020, fracking is projected to lead to between 200 and 800 extra premature deaths each year from ozone and particle pollution across the Marcellus and Utica Shales in the Appalachian Basin, which have seen intensive leasing and drilling. This large area of the eastern United States crosses several states.[33] Fracking fever has swept through the most densely populated parts of Europe,

* At monitoring stations on Cape Verde, the Canary Islands and Iceland.

including Denmark, Lithuania, Romania and especially Poland.[34] With an increasing reliance on imported gas from Russia, and the desire to meet carbon emission targets, the pressure for shale gas exploitation will continue. Maybe there is a silver lining: if shale gas leads to the shutdown of polluting coal-burning industry and power plants or the displacement of extensive oil heating in cities such as New York, it could help urban air pollution. This, however, should not be used as a reason for poor controls on the natural gas industry.

One of the frequent justifications for fracking is the use of natural gas as a bridging fuel between coal and a low-carbon future. Certainly, burning natural gas for energy emits a lot less CO_2 when compared with burning coal. However, natural gas is mostly methane, which has strong global warming effects in its own right. Natural gas therefore only provides climate benefits over coal if leakage is tightly controlled, to no more than 2 to 3 percent. Leakage rates of at least 0.18 to 2.8 percent have been measured in flights over fracking wells even before the gas is distributed to users.[35] It will therefore be hard for US fracked gas to be better for the climate than burning oil or coal.

So why was agreement reached on the chemicals that destroy ozone in the stratosphere, while international collaboration on ground-level ozone has largely failed? Controls on ozone in the stratosphere required bans on chemicals used in refrigerators, aerosols and fire-fighting around the world. These chemicals were made by a small number of companies and near like-for-like replacements were available. We were not being asked to change our lifestyles, just make some adjustments to our technologies. Controlling ground-level ozone will require us to rethink the way in which we use petrochemicals and natural gas, and the problem extends to the way in which we manage land. This will be much harder. But without international action, ozone at ground level will cause ever-increasing harm to our health and damage to our crops.

Clearly there are huge challenges before us to reduce the intolerable health burden from air pollution and they will be greatest in those countries that have the least resources to tackle them. The worsening air quality in India and surrounding countries requires urgent action. Some of the benefits of development, urbanization and industrialization will be counteracted by rising death rates if air pollution strategies are not integrated into global economic development plans.

Growing urbanization poses additional challenges and opportunities. In 2015, for the first time in human history, over half the world's population lived in cities. However, in 2015, only 12 percent of urban dwellers enjoyed air quality that met WHO guidelines. Half of the world's megacities had air pollution that exceeded guidelines by two and a half times, and in most places it is getting worse.[36] Even in the wealthiest parts of the world, across Europe and North America, it is not clear that urban air pollution is decreasing. The current emphasis on technical strategies to clean up our air is not working, and in some cases, progress that has been made is being undone by other trends—for example, Europe's increased use of diesel cars and wood-burning (see Chapters 10 and 11). Growing urbanization is leading to a growing world health problem. More than ever we need to transform existing cities through design, reducing the growing dependency on road transportation and providing clean home energy. New cities need to be sustainable, with low energy use, reduced transportation use and low pollution. This includes investing for the nearly one billion urban poor who live in informal settlements so that they can be near the economic opportunities that cities bring. Once a city is built, it is hard to change its physical form and land use. These could be locked in for centuries. We will pay a heavy price if we get it wrong now.

Chapter 9

Counting particles and the enigma of modern air pollution

In 1996, Scottish scientist Anthony Seaton was thinking about the enigma of modern air pollution.[1] Coal-burning in UK cities was under control and our particle pollution was lower than it had been for centuries, but the Six Cities study had just told us that people were dying early because of particle pollution. There were problems closer to home too. London had just had its first winter smog caused by modern traffic pollution, and between 101 and 178 people had died.[2]

Seaton was a medic. He had worked as a chest physician in south Wales and in the mid-1990s he was the director of the Institute for Occupational Medicine in Edinburgh.[3] Formerly a government laboratory, the institute carried out research on workers in dusty environments such as coal mines and factories. Herein lay the first part of the puzzle. Workers in many factories were breathing air pollution hundreds or even sometimes a thousand times greater than the air particle pollution that they breathed outside, but they remained, largely, fit and healthy.* However, the air pollution that people breathed outside in towns and cities was shortening their lives. It did not make sense. It was equally hard to understand how air pollution killed people. The people who died prematurely did not just die of lung disease. Dockery's Six Cities team had recently found that people also died of heart attacks and strokes.

Standards and regulations on the quality of the air that we breathe have always been based on the mass of the particles, normally measured as micrograms per cubic meter. We buy food or trade goods by the ounce, gram, pound, kilo or ton, so it makes sense to think of the amount of a substance in terms of its weight.† But how can such tiny concentrations of particles harm us so greatly when we evolved in naturally dusty environments on the African continent? This has puzzled toxicologists for many years. It isn't just a case of more particles in our air. Clearly something must be different about the air pollution that we are exposed to today, compared with the environments that we inhabited as *Homo sapiens* evolved.

The chemical composition of modern pollution is of course one huge difference. We breathe pollutants that our ancestors never met. Another difference is the number and size of particles in our urban air compared to those found in natural environments. We have always been exposed to particles, but these have been relatively large

* One obvious solution to this conundrum is the healthy worker syndrome. Workers are the healthiest people in society. Young children, the elderly and the ill do not work in factories but they do breathe city air. However, even this did not stack up. Outdoor air pollution was harming everyone, not just the most vulnerable.

† The SI unit for amount of a substance is the mole, but in everyday life we simply weigh things.

particles of soil dust, pollen or sea salt, which are trapped in our nose and throat. The particles in modern air pollution are much smaller and, for this reason, they can be breathed deep into our lungs.

In 1996, Seaton received some new measurements from Roy Harrison of the University of Birmingham. These showed that the number of particles in modern city air could rise to over 100,000 in each cubic centimeter that we breathe. The sheer number of particles that we breathe is astonishing. Stand in the middle of a city park and each time you inhale you will be taking in about two million particles—twenty million if you walk along a busy road or stand near an airport perimeter fence.

Seaton put forward a new idea. It was not the mass of the particles that was important but their number. Here, size matters a great deal. Buy a kilo of apples and you will get about ten or twelve. Buy a kilo of rice and you will get about forty to fifty thousand grains. Similarly, a single large pollen or dust particle can weigh the same as tens of thousands of particles emitted from a car or aircraft exhaust.

Seaton understood what happened to particles when we inhaled them. Around half of the tiny particles would be breathed in and straight back out again, but the remainder would be deposited in our lungs. Here the numbers matter a lot. Let's think again about our kilo of apples and kilo of rice. Drop your bag of apples on the kitchen floor and they will cover just a small area and can be easily picked up. Accidentally split your bag of rice and it will scatter everywhere, covering the kitchen floor; cleaning up will be a big hassle. Similarly, when we breathe big natural particles such as pollen or dust they will be deposited in our lungs in a small number of locations. Our body's defenses will be mobilized and the particles will be removed.

However, the millions of tiny particles in each breath of modern air are scattered across the whole of our lung surface. In an adult that can be around the same area as half a tennis court, meaning that our body's defenses have to mount a massive cleanup operation. Seaton suggested that the inflammation caused by this cleanup would activate the body's immune systems, add to blood clotting and therefore increase the risk of heart attacks and strokes.

Soon after Seaton published his theory, concerns about the health effects of breathing very small particles was catapulted into the mainstream media for another reason: the debate on the safety of nanotechnology. Today we take self-cleaning windows, effective suntan lotions, better paints, medicines, high-capacity batteries, ever faster computer processors and cellphone screens for granted. However, around the turn of the century, engineering and creating the very tiny particles that are embedded in these technologies was highly controversial. What would happen if we breathed our spray-on suntan lotion, and what happens when it washes down the drain? Since the benefits are more obvious today, there is less public debate about the perceived risks from nanotechnology and we use these products glibly. In the early 2000s, however, scientific advances were causing fears: first due to the growth of genetically modified plants and second thanks to the unknown nature of nanotechnology and the terrifying idea that it would lead to tiny, self-replicating nanorobots that would one day multiply uncontrollably like viruses and devour our planet. Even the Prince of Wales joined in the debate. In fairness to Prince Charles, he tried to draw attention to the risks and promote a reasoned dialogue. But this did not go in the intended direction; instead it produced headlines about

out-of-control nanotechnology polluting our environment such as the infamous "Prince fears grey goo nightmare," describing an end state feared by some environmentalists from out-of-control, self-replicating robots. The Prince denies that he ever said the words.* The UK's Royal Society was prompted to investigate the environmental and health risks posed by new nanoparticles and nanotechnologies in all their forms. Their comprehensive report in 2005 focused on manufactured nanoparticles but it also called for greater research into the nanosized particles that we breathe every day from vehicle exhausts and other sources.[4]

Despite these recommendations and the concerns of scientists, there has been relatively little research into the health effects of the number of particles that we breathe. With the Seaton hypothesis originating in the UK, and the debates prompted by the Prince and the Royal Society, concerns about breathing vast numbers of particles had more leverage with policymakers there than in other parts of the world. Routine counts of particles in urban air began in the early 2000s. By 2005 there were enough measurements to be able to compare them to health statistics. I was part of a London-based team, led by Richard Atkinson from St. George's Medical School, that gathered data on daily deaths and hospital admissions in the capital and compared them to the number of particles in the air.[5] Many things affect these health statistics, including temperature and access to medical care on certain days of the week. Having removed these factors, what we found really surprised us. When the *mass*

* The "grey goo" quote appears to have been wrongly attributed to the Prince; see his 2004 speech at https://www.princeofwales.gov.uk/media/speeches/article-hrh-the-prince-of-wales-nanotechnology-the-independent-sunday. See also https://www.telegraph.co.uk/news/uknews/1431995/Prince-asks-scientists-to-look-into-grey-goo.html and the Michael Crichton novel *Prey*.

of particles in the air grew, more people died or were admitted to hospitals with respiratory problems, but on days when the *number* of particles increased, there were more heart attacks. The larger particles that we can weigh might be causing breathing problems, but smaller nanosized particles that we can only measure by counting them might be causing heart attacks. This was worrying. The mass of particles in the air was clearly improving, but the number of particles that we measured were much the same as John Aitken had found with his portable cloud chamber and microscope a hundred years before.

It is rare to find a positive news story about urban air pollution, but in late 2007 something dramatic happened to the number of particles in the air across the UK. There is an old adage in air pollution science concerning the difference between the people who predict air pollution using models and those (like me) who measure it. It runs like this: no one believes the results from predictive computer models other than the modelers who make them, and everyone believes the measurements apart from the people who run the instruments. So, true to form, when the number of particles in the air alongside London's Marylebone Road dropped by nearly 60 percent in just a couple of months[6] we thought the instrument had malfunctioned. Even when we found big decreases at the same time at a central London school and in Birmingham, we thought it was a fault common to all the instruments.

There was no fault. At the end of 2007 the UK took the final step in introducing ultra-low sulfur diesel. The maximum amount of sulfur allowed dropped to 0.001 percent, from about 0.003 percent. This tiny extra change had a dramatic and completely unexpected effect. I cannot think of another policy that has improved air quality so dramatically or so quickly. However, the improvement was not part of a government plan; it was an unintended consequence. The sulfur in diesel was reduced to allow new technologies to be fitted to diesel exhausts. The UK was in fact one of the slower European countries to mandate ultra-low sulfur diesel. A similar fall in particle numbers occurred when ultra-low sulfur diesel was introduced in Denmark in 2006. Still, the number of particles in the air across the UK has continued to decline in line with tighter exhaust standards on new diesel vehicles.

Despite the early work of Aitken, we are only now taking baby steps in understanding the sources of these tiny particles in the air of our cities. Governments and city measurement programs focus on the pollutants that are harmful. These are subject to regulation and laws, but a chicken and egg situation arises when a new pollutant emerges. Without sufficient measurements, studies cannot be done to look at health impacts in the population and governments are reluctant to make measurements until the health risks can be proven. We were able to identify the problem resulting from the presence of sulfur in diesel, but other sulfur-rich fuels are culprits too: most notably, aviation kerosene.

Look up into the sky over Europe or North America and you will almost certainly see contrails. These are not smoke from the engines. Contrails typically form a few wingspans behind the plane and you cannot see them as a passenger. As in Aitken's particle counter, tiny ice crystals form around the particles in the aircraft engine exhaust, making them visible. Contrails have been studied since the 1940s but comparatively little research has been done on the number of particles emitted by aircraft when they are on the ground, where we live and breathe. Airports represent a really challenging environment for aircraft engines. It is here that they have to start up, and then produce low thrust for taxiing followed by maximum power at takeoff, but they are optimized for high-altitude cruising where most fuel is used. The concentration of particles in the air at an airport is huge. At the perimeter fence, hundreds of meters from the runways, the number of particles can be about the same as those found at the curb of a busy London street, just a couple of meters from the traffic.

Los Angeles is an ideal place to study how airport pollution travels across a city. The international airport is on the coast and a persistent onshore breeze carries pollution inland. In 2013, scientists from the University of Southern California loaded equipment onto a hybrid electric car and drove around the airport. It was clean on the upwind side, but huge numbers of particles were found at the downwind fence. Next, they drove away from the airport in a zigzag pattern along the city's gridded streets, following the flight paths. Even eleven miles from the airport they were still able to find particles from aircraft, and these were increasing the number of particles in the city air by more than ten times.[7]

Problems with particles around airports are not confined to LA. In 2012, I met Menno Keuken at a conference in Brussels. We have a shared interest in air pollution measurement and he asked me to look at some new data that he was having trouble believing. Keuken was taking measurements in the middle of the Dutch countryside, near the small village of Cabauw in a mainly farming area. The village is home to around seven hundred people, with some farms located along the canal. It is also home to a 213-meter-high tower nicknamed *de snuffelpaal* (the sniffing pole).* This rises above the flat Dutch terrain and allows scientists to measure air pollution at various heights. Air pollution found here can be tracked back in one direction to the port of Rotterdam and in another to the industrialized areas of the Ruhr in Germany.[8] In 2012, they started counting particles and discovered a new pollution source they had never seen before. It lay to the northwest. A line was drawn on a map and followed. There was very little industry in this direction. It was mostly farmland. Twenty-five miles away the line reached Schiphol, Europe's third largest airport. Menno could not believe that he had detected the presence of the airport over such a large distance. He wanted to double-check, so he set up a measurement site about four miles from the airport at Adamse Bos, a parkland area on the outskirts of Amsterdam and largely away from the flight path. When the wind blew from the city he counted about 14,000 particles in each cubic centimeter of air. When it blew from the airport he counted over 42,000 particles, an astonishingly high number given the distance from the source. Around 200,000 Dutch homes were being exposed to these airport particles.

Airport expansion has been a fiercely debated topic in the densely populated southeast of England. Despite extensive and controversial impact assessments and government inquiries, the possible impact of breathing the huge numbers of particles that come from airports has not been considered. However, one study does raise an interesting question. We have already met Anna Hansell and her team at

* For a history of the Cabauw observatory and the sniffing pole, see http://www.cesar-observatory.nl/cabauw40/index.php.

Imperial College London. Hansell is also interested in aircraft noise and its effects on the 3.6 million people living around Heathrow Airport.[9] She found that those experiencing the greatest aircraft noise had more strokes and heart attacks, but the patterns did not quite follow the flight paths. The patterns were consistent with air pollution extending from the airport. Could this be a particle number effect?

Why do aircraft produce so many particles? The black smokc that trailed behind Concordes and the aircraft of the 1970s is a thing of the past. Today, aircraft engines are much quieter, cause less air pollution and use less fuel. But the answer to the particle number question is not in the engine. It is in the fuel. In the United States and Europe, sulfur is removed from the diesel and gasoline that we use in our cars, buses and trucks. It is not removed from aviation kerosene. The maximum amount of sulfur in aviation fuel is three hundred times greater than in road fuel. Not all aircraft fuel is this bad—typically it has about sixty or seventy times more—but it is sulfur in the exhaust that spontaneously forms huge numbers of tiny particles.

The answer to the aircraft particle problem seems straightforward: simply remove sulfur from the kerosene as we do with road fuel. However, there are a couple of stumbling blocks. Low-sulfur aviation fuel lacks the lubricating and anti-corrosion properties of current aviation fuel and, most importantly, the aviation industry consumes vast amounts of fuel—and removing sulfur costs money. Setting aside the possible health impacts from the number of particles, removing just the sulfur from fuel used by cruising aircraft has been estimated to prevent around 900 to 4,000 early deaths per year.[10]

It is perhaps not so surprising that large numbers of particles come from aircraft, traffic and industry such as oil refineries, but street surveys have found another culprit: fast-food restaurants. In 2010, scientists in Vancouver were trying to create an air pollution map of their city.[11] For three weeks they went out every day with handheld instruments and stood at eighty survey points around the city. They found air pollution from traffic, as you would expect, but a new finding surprised them. The distance of the survey point from a fast-food restaurant was a big factor in the number of particles that they found; specifically, there were more particles in the air when there was a fast-food restaurant within about two hundred meters.

Researchers in Utrecht took up the challenge to investigate this phenomenon.[12] For three weeks, Cristina Vert from the city's university walked a set route around the city center during lunchtime and in the evening, pausing for a few minutes outside each restaurant before doing a circuit of the city square and crossing the canal. A quick glance at Utrecht on Google Maps shows a huge number of bars, restaurants and cafés in which to while away the evening. Vert investigated seventeen of them. Most used frying or grilling. Passing mopeds and—reminiscent of Aitken's measurements near a Bunsen flame—outdoor candles also added to the particle number, but restaurants were the biggest source. Seaton and his colleagues had found high pollution and large numbers of particles in indoor environments about fifteen years earlier, but finding that cooking had an impact on outdoor air was a surprise.* Clearly catering kitchens would be unworkable environments without extractor fans, but Vert's work showed that indoor air pollution can affect the outside.

* Fat particles from cooking have also been found in city air, including that of central London.

Many cities have pollution maps for the classical pollutants, such as nitrogen dioxide, ozone or the mass of particle pollution in our air, but models for the number of particles have a long way to go. This must start with a better understanding of the sources, but the behavior of particles in the air makes mapping hard. For instance, particles can stick together. This does not affect the mass of the particles in the air, but it does change their number. Another difficulty is that new particles can form spontaneously on sunny days in largely clean air. Why, or how, is still largely a mystery. These events used to be rare in cities. They were almost unknown in northern Europe, although they were seen occasionally in southern Europe where the sun is stronger. But, as urban air quality improves, these events are becoming more frequent. Sometimes they can be confined to city centers, but at other times a whole region can be affected for many hours by pollutant gases coming together to form large numbers of particles.[13] In 2011 and 2012, these spontaneous events created around 12 percent of the particles that were counted in London's air.[14] The formation of new particles in relatively clean air makes it difficult to predict the number of particles that we breathe and more difficult still to control them. We have a long way to go before we understand the particles that Aitken began to count over one hundred years ago.

Chapter 10

VW and the tricky problem with diesel

In September 2015, air pollution made world headline news as never before, when the German car manufacturer Volkswagen (VW) admitted to cheating exhaust emission tests.

Cars sold in Europe, the United States and most other parts of the world have to pass air pollution tests before governments will approve them for sale. In many ways this is no different to road safety requirements. Software had been found in some VW cars that could recognize when the car was being tested. The onboard computer would then adjust the engine and exhaust controls from their normal mode in order to pass. VW's chief executive resigned and faced prosecution, while a process imposing financial penalties is still ongoing. A contagious loss of confidence in diesel cars slowly began. By 2017, diesels had dropped from around 50 percent of new car sales to around 35 percent in Germany and the UK.

The VW scandal was exposed by the International Council on Clean Transportation (ICCT), a nonprofit organization that provides technical and scientific advice on low-pollution transportation.* In 2013, the ICCT was looking at the exhaust performance of diesel cars sold in the United States and found that they emitted far more nitrogen oxides in real-world use than they did in the test. The data was passed to US regulators and VW was issued a violation notice in September 2015. The company soon admitted that around 500,000 cars sold in the United States had been equipped with illegal software to detect if the car was being tested and lower its emissions accordingly.[1] The scandal then spread to include 8.5 million cars sold in Europe and 11 million worldwide.

Although the VW scandal initially broke in the United States, the number of diesel cars in the States is very small. Less than 5 percent of cars are diesel powered; most cars have gasoline engines and diesel is largely a fuel for trucks and buses. By contrast, Europe embraced diesel cars. Millions were sold and by 2015 around half of the cars on European roads were diesels. Given Europe's relatively small size it seems astonishing that around 70 percent of all the world's diesel vehicles are on European roads.[2]

The VW scandal was a shock to investors, car owners and governments alike. There was also huge surprise among air pollution scientists (like me) who had spent their careers trying to understand air pollution in European cities. It turns out that VW was just the tip of the iceberg. In 2016, the European Parliament's inquiry into the VW scandal concluded that many other vehicle manufacturers had exhaust control strategies that were "irrational" and "unjustified by technical limitations," and that "some car manufacturers have opted

* The ICCT is an independent organization founded in 2001 to provide unbiased research and technical and scientific analysis to environmental regulators.

to use technology that assures compliance with emission limits only in laboratory test, not for technical reasons but for economic reasons."[3]

To unpack the ramifications, we need to understand why diesel cars became so popular in Europe and how scientists recognized that there was a lot more going on than just the illegal software in some VW engines.

In Europe, diesels have been marketed as a low-carbon mode of transport and better for climate change when compared with gasoline vehicles. This notion was embraced with surprising enthusiasm in the form of tax breaks for diesel in every European country. Given politicians' reluctance to create tax incentives to mitigate air pollution, it seems amazing that the assertions about diesel's low-carbon advantages were subject to almost no challenge from governments across Europe. At 2017 rates, the average European diesel car owner received a 2,000 euro subsidy through tax breaks over the lifetime of their car compared with the owner of a gasoline car.[4*] The EU countries with the lowest diesel taxes were generally home to the biggest car factories, or were geographically well placed to sell large volumes of diesel (and therefore yield tax) to the vast volume of international freight traffic that travels on Europe's roads.[5]

It is rare that a political economist and an environmental chemistry professor get together to study an issue, but this was exactly what was required to uncover Europe's diesel issues. They were not just technical or engineering problems, but were created over

* One possible explanation for these historic differences could be that governments have sought to maximize the income from fuel taxes but were wary about pushing up the cost to business of delivering goods and services. Most haulage and freight movement by road has been diesel powered while most private motoring used gasoline, and hence it was economically easier to tax gasoline. The UK is unique within Europe in taxing both fuels at the same rate per liter.

decades by political and economic decisions. In 2013, Michel Cames from the University of Luxembourg and Eckard Helmers from Trier University of Applied Sciences in Germany were the first to question the accepted wisdom that diesel was better for the climate than gasoline. They suggested that reducing carbon emissions was a smokescreen that European governments were hiding behind. In reality the push to embrace diesel was economic, not environmental. Their evidence trail was published not in a newspaper or an online blog but in a peer-reviewed scientific journal.[6]

Problems began in the late 1960s, when European natural gas fields first came online. At this time, a lot of heating oil was burned in boilers to warm schools, offices and factories, as well as some homes. It was clear to the oil companies and governments that natural gas would replace oil and the market for heating oil in urban areas was largely finished. Many power stations too were oil-fired,* and the drive toward nuclear power, especially in France, meant that oil would also be displaced from that market.

The crude oil that comes from the ground is not a single product but a mixture of oils that are separated in oil refineries for different uses. The lighter fractions are used for gasoline and the heavy fractions are used in shipping, but the advent of natural gas in Europe meant that there would be no market for the middle fractions that had been used for heat and electricity.

What would the oil companies do with a large part of their production? According to Cames and Helmers, the European diesel car boom had nothing to do with reducing climate impacts, as we have

* London's Bankside B power station, now home to the Tate Modern, was oil-fired. You can walk inside the old storage tanks, which have been converted into gallery space.

been frequently told. It began long before climate change became an acknowledged issue in the late 1980s. The promotion of diesel cars came from the oil companies, government and vehicle manufacturers, who worked together to find a new market for the middle fractions of crude oil. If they could not be used in heating and power generation, then diverting them into the road fuel market was the only answer. Trucks and buses already ran on diesel, so to consume more we had to start using diesel in our cars and small vans, which had been previously powered by gasoline. The strategy was formalized with the European Auto-Oil Programmes throughout the 1990s. European car manufacturers invested in diesel engine technologies, the drivability of diesel cars improved, and taxes encouraged us to buy them. The oil companies developed a long-term market for the middle fractions of crude, the car companies sold diesel cars and, thanks to tax breaks, driving became cheaper. European car manufacturers would also become global leaders in building small diesel engines, giving them a technology to export around the world.

It seemed like wins all around. Later, with climate emission targets needed across all sectors, European car manufacturers, government policies and tax regimes were further constructed to favor diesel as an apparently lower carbon emission fuel. So, did it work?

Certificates displayed in car showrooms certainly showed that the carbon dioxide emissions from new diesel cars were lower. And as more diesel cars were sold, it seemed obvious to everyone that the policy was working. The better miles per gallon or kilometers per liter added to the perception of diesel as a low-carbon fuel. But this ignored the simple fact that a gallon of diesel contains a lot more energy and releases a lot more carbon dioxide than a gallon of gasoline. If diesel and gasoline were taxed on the energy content of the fuel rather than volume, then diesel taxes would be around 20 percent higher than gasoline.*[7] However, car owners began to notice that the fuel economy that they achieved once they owned the car did not match what they expected from the certificate in the showroom. In 2000, drivers were finding that diesel consumption was about 8 percent worse than advertised. By 2013 it was 38 percent worse.† The gap was bigger for diesel cars than gasoline ones, meaning that in real-world use their carbon dioxide emissions differed by just a few percent.

There was another side effect that further eroded the perceived climate benefits of diesel cars; their exhaust also contains sooty black carbon particulate, whereas gasoline engines emit almost none. In addition to their impacts on our health, black carbon particles are important for climate change since they are strong absorbers of the sun's heat. Due to Europe's geographical position and a mainly southwesterly air flow, it is the black carbon from Europe that dominates the soot deposited on Arctic snow, where it has a climate warming effect and encourages snow melt.

Even for the oil companies, it is possible for a plan to be too successful. By the end of the first decade of the twenty-first century, the shift to diesel cars had gone too far and demand for diesel exceeded what Europe's refineries could produce. A rational policy response would have been to increase diesel taxes, but that did not

* The EU Commission has proposed equal energy rather than volume-based fuel tax but member states, each answerable to their electorates and car manufacturers, have resisted.

† The reasons for the widening gap are largely unknown but are thought to have come from manufacturers gradually getting better at pushing the boundaries of the test, including removing wing mirrors and taping over all gaps in test vehicles to make them more aerodynamic, and running them with completely smooth tires with no tread. See Kühlwein, J., *The Impact of Official versus Real World Load Tests on CO2 Emissions and Fuel Consumption of European Passenger Cars.* Berlin: ICCT, 2016.

happen. Instead, Europe started importing diesel made elsewhere and exporting gasoline that our oil companies could not sell. So where did the extra diesel come from? Cames and Helmers traced the supply chains for much of this extra diesel back to refineries in Russia. These were old and inefficient, meaning that they used a lot of energy, and therefore produced a lot of carbon dioxide, to make each gallon of diesel, further denting its climate benefits. So, adding together the real-world fuel economy, the black carbon and the extra emissions from Russian refineries, the European diesel boom from the late 1990s to the time of the VW scandal had no benefits for global climate, contrary to the accepted wisdom.

The natural gas discoveries that changed Europe's use of oil were not replicated in the United States, Japan and other major car markets. Even well into the twenty-first century, many of New York's iconic skyscrapers are heated by oil, delivered by tanker. Japanese manufacturers choose to invest in gasoline technologies rather than diesel. Over fifteen years, from the mid-1990s to the early 2000s, the carbon dioxide emissions from Japan's new cars fell further, and faster, than Europe's and by 2010 the average Japanese new car emitted less carbon dioxide than the new diesels being sold in Europe. By investing in diesel car technologies, the Europeans had backed technologies that no one else needed. It is the desperate efforts of European car manufacturers trying to break into the US market that set the scene for the VW scandal to break.

To make sure that diesel uptake was not hindered by tough pollution limits, the European exhaust standards allowed diesel cars to emit more particle pollution and nitrogen oxides than gasoline cars. The difference was not small and completely ignored health impacts.

The standard imposed for European gasoline cars sold between 2000 and 2005 was three times tighter than comparable diesels. In the United States, with little indigenous diesel car manufacturing, there was no reason for the US Environmental Protection Agency to favor diesel in the same way as European regulators, so both gasoline and diesel vehicles had to meet the same exhaust standards. The VW scandal broke when the ICCT were trying to understand how the same car could be sold in both Europe and the United States and yet meet two different sets of standards. It was VW's efforts to answer this question that led to their admission that illegal software had been installed that would modify the car's performance during a test.

The VW scandal is not directly about climate change or particle pollution. It revolves around nitrogen oxides,* a family of pollutants that includes nitrogen dioxide. To understand the problem, we have to go back to 1999, when the European Union (EU) followed the World Health Organization guidelines and set a limit for nitrogen dioxide emissions. The limit had to be met eleven years later, in 2010, giving everyone plenty of time. Or so they thought.

The UK government's initial response to the EU's 2010 limit was not to try to wriggle out of it. Instead, the UK set itself a more ambitious target. In 1999, the UK decided to meet the EU limit five years earlier, by 2005 not 2010. But it quickly became apparent that this

* Nitrogen oxides are produced in hot combustion. In the main they do not come from the fuel itself but from the hot burn conditions, when oxygen and nitrogen in the air combine together. The result is mainly nitric oxide gas, a molecule of one nitrogen and one oxygen atom. Most combustion—vehicles, gas heating, power stations—produces nitric oxide. Tiny amounts of nitrogen dioxide (one nitrogen and two oxygen atoms) are produced too. This is the pollutant of health concern. Nitric oxide also becomes nitrogen dioxide as it slowly mixes with oxygen once the vehicle exhaust or factory plume meets the fresher air. Controlling nitrogen dioxide therefore requires control of both the nitric oxide and the nitrogen dioxide emitted into our air.

challenge was going to be harder than expected. I was working as an air pollution scientist in London in 2001. My office looked out onto the Houses of Parliament and Westminster Bridge, which was a great position to ponder the traffic. Our team at King's College London quickly realized that, with just a few years to go, nitrogen oxides in the exhausts of all of those vehicles crossing the bridge and driving throughout London would have to be cut in half. More curiously, at that time, we measured a lot more nitrogen dioxide in central London than expected.[8] All sorts of theories were put forward, including that there was something special about the exhausts of black cabs or buses that filled the streets of central London, but in truth we had no idea.

However, no matter how bad the situation was, we knew that things could only get better. Ever tighter exhaust standards were being applied to new vehicles, meaning nitrogen oxides from cars and vans were going to be regulated for the first time.* From 2005 on new cars would have to pass stricter laboratory tests before they could be approved for sale. Exhaust limits would be halved. So London would not meet the target by 2005, but we believed that tighter laboratory tests meant an inevitable reduction in pollution on our streets. We had little doubt that London's nitrogen dioxide problems would be solved by 2010.†

It was not to be. Instead of getting better, the situation became worse. Sitting opposite London's Madame Tussauds' waxworks is the most advanced urban air pollution facility in Europe. It is about

* For information on standards and tests see https://dieselnet.com/.

† In 2001, our team had a series of meetings about what to do when the air pollution problem was sorted in just a few short years. Fearing for our jobs, we hired a noise specialist to diversify the group. We need not have been so concerned.

the size of two shipping containers and bristles with tubes and pipes that sample the air right next to the busy Marylebone Road. Behind steel doors, inside windowless cabins, the noise of the passing traffic is drowned out by the rattle from pumps, the percussion sounds of valves and the hissing noises of air rushing through pipes as instruments measure what Angus Smith would have described as impurities in the air. Sometimes, when I am there on my own, I switch off the lights and stand still in the main analysis section, marveling as the laboratory is lit only by the numbers tumbling across the illuminated instrument displays and screens. Over 80,000 vehicles drive past the facility every day. It was installed and is managed by my colleague and friend David Green, but scientists from around the world use the data. It has been the venue for many air pollution discoveries since it was first craned onto the pavement in 1997.

One of these discoveries forms part of our diesel story. In 2003, in just fifteen months, instead of decreasing, the level of nitrogen dioxide alongside London's Marylebone Road increased by over 25 percent. At first it was dismissed as a blip. Being good measurement scientists, Green and I set about checking the result and installed another instrument to measure nitrogen dioxide, but both instruments told us the same thing. We hoped that it was a local effect, but that hope soon evaporated. Around London, and indeed the UK, it soon became clear that concentrations of nitrogen dioxide were going up, not down. An international air pollution disaster was slowly unfolding. The best guess was that technologies fitted to diesel vehicles to clean up other pollutants were making the nitrogen dioxide problem worse. One thing we did realize was that the European standards had left something out. Controlling the whole family of nitrogen oxides in exhausts (as the Euro standards did) was not the best way to control nitrogen dioxide. It needed to be targeted specifically, and this was missing.[9]

In what was to become a recurring pattern, government placed its faith in the next round of tighter exhaust standards, due in 2005, and importantly they believed the assurances of the car makers. There was no need to change the policy or look closer at vehicle exhausts; we just had to wait a little longer. It was like the promise of jam tomorrow in Lewis Carroll's *Through the Looking Glass.* As the new vehicles hit the roads, pollution by nitrogen dioxide and indeed the whole family of nitrogen oxides in our cities got worse, not better. Between 2005 and 2010, nitrogen dioxide from traffic in London and in Paris increased by an average of 5 percent a year.[10]*

By the time the VW scandal broke in 2015, European cities were not even close to attaining the legal limits. Alongside some London streets, the level of nitrogen dioxide was around three times the amount the law allowed. It was a shameful failure of policy to control air pollution and prevent thousands of early deaths. In the UK alone, it was estimated that around 23,500 early deaths per year were due to the nitrogen dioxide that we breathed.[11]† Although air pollution scientists were surprised that a car company had been cheating, it was clear as soon as the VW scandal broke that the issue of nitrogen oxides from diesel vehicles was far bigger than just one

* Greece was a notable exception to the European trends. Diesel cars were rare on Greek roads due to restrictions on sale and use, making the fleet very gasoline dominated. Here a strong downward trend in nitrogen dioxide was seen from the mid-1990s and into the early 2000s.

† This is considered a high estimate. Given the overlap between NO_2 and PM exposure in health studies, the effects of NO_2 alone are probably less than this. The Royal College of Physicians gave an estimate of 25 percent less. See Royal College of Physicians and Royal College of Paediatrics and Child Health, *Every Breath We Take.*

car maker. VW alone could not account for such a widespread failure of policy. The problem, whatever was causing it, was endemic in the vehicle fleet. So, what went wrong?

The first suggestion in the run-up to the 2010 deadline was simply the growth in the number of diesel cars and vans on our roads and their looser exhaust limits when compared with gasoline. However, growth in the dieselization of the vehicle fleet should not have offset the tightening of exhaust standards.

What was really in the exhaust of the vehicles on our roads? A team led by David Carslaw set out to investigate.[12] Instead of testing cars one by one, they measured tens of thousands. Specialist equipment was brought in from the University of Denver. It was parked alongside roads in London, and light beams were shone into the exhaust of each passing vehicle. The equipment was the invention of Don Stedman, and everywhere his equipment went, Stedman went as well.* Often his wife came too. I spent one summer's afternoon with Carslaw, Stedman and Stedman's wife measuring traffic exhaust on London's Queen Victoria Street. I cycled there from our operations center at King's, folded my bike and put it among the masses of wires and equipment that hung from the back of a small white van. Stedman and his wife were well prepared for these long days in the field and were comfortable in their deckchairs. I brought tea from a nearby café and was allowed the honor of being able to sit inside the parked van. About ten seconds after each car passed through the light beam, a measurement would briefly appear on a rather old laptop in front of me and I would call out the result. After

* Sadly missed: https://magazine.du.edu/campus-community/chemistry-professor-donald-stedman-dies-lung-cancer/.

a while I became pretty good at looking at the passing car, seeing its make and model and guessing the result before it appeared. But Stedman was brilliant and spotted the most polluting vehicles in the distance, before they reached us. He was right every time.

Stedman mainly worked in the United States, where nearly all cars were gasoline. All afternoon his excitement and amazement at the exhaust produced by each diesel car did not wane. As the working day drew to a close, the traffic gave us a special treat as the well-paid directors of banks in the City left their offices in cars that were just weeks old, fresh from the showrooms and tested to the tightest standards that Europe had seen. Stedman let out a whoop with every result. It was the newest diesel cars and vans that were producing the most nitrogen dioxide. Three-way catalytic converters on gasoline cars were working well, even better than expected. Sometimes when a new gasoline car would pass through the light beam we could detect nothing at all. The new gasoline cars were emitting a fraction of the nitrogen oxides that came from gasoline cars made ten years before, but for diesel vehicles it was a different matter. Despite ever-tighter testing standards, the exhaust from newer cars was no better than that from older ones and often much worse. So why—and even more importantly, how—were new vehicles passing ever tighter tests but not producing less pollution in the real world?

During the official tests, the manufacturer would place each new type of car and van on rollers in a laboratory and drive it through a set cycle. The vehicles would gently accelerate up to city driving speeds and then slow down again, followed by a simulated six-minute drive between two cities. This is nothing like the driving that you or I do in real life.

It is rare for me to meet a vehicle engineer. The conferences and scientific meetings held for environmental scientists do not seem to attract vehicle engineers, and I think this is part of the reason for the diesel exhaust scandal. Soon after my time with the Stedmans, I had a phone call from Lionel Moulin, who worked in the French environment ministry. A big transportation technology conference was going to take place next to his office in Paris. He felt that someone needed to speak on air pollution. His solution was to propose a conference session on air pollution from vehicles, but he needed some help.

It took two attempts, but finally the organizers agreed to include our air pollution workshop, though they scheduled us at five o'clock, at the end of the day. I arrived early in Paris, bounded off my Eurostar train and set out to absorb some of the science and to tour the conference exhibitions. It was nothing like air pollution science events. There was money there, and lots of it. It was full of thousands of people—industry trying to sell to industry—and it even had the latest concept cars, curiously adorned with young, sleekly dressed models giving out leaflets.

The speakers met in the green room beforehand. We planned our session and at the appointed time we walked onto the stage. It was a big auditorium that could hold five or six hundred people. It was also the only time I have ever spoken when there were more people on the stage than in the audience. No one wanted to know about the inconvenient issues around air pollution. For this reason, the discussion session did not last as long as we planned, and we finished early. I took the opportunity to talk to a French vehicle engineer. I asked about the exhaust tests that vehicles have to pass and why they are so different from driving in the real world. He grinned from ear to ear, shrugged and explained it was because the test was old. It dated from around 1990 and was designed around the cars and vans being sold at that time. These included the Citroën 2CV, a vehicle first produced in the 1930s to motorize the many French farmers who were then using horse-drawn carts.[13] Designed to carry four farmers and 50 kg of potatoes to market at a top speed of 60 km/h (about 40 mph) and to be able to cross a field carrying a basket of eggs, the 2CV became a style icon in the 1970s, much loved by hippies and ecologists, but it was not a performance vehicle. The final and fastest production vehicles made in the 1990s had a theoretical top speed of 155 km/h (70 mph) with the acceleration to match. So, every time a car was driven outside the performance envelope of a 2CV (which was a lot!) it fell outside the test criteria.

Throughout the first decade of the twenty-first century, diesel cars were fitted with ever more powerful engines to make them drive like performance gasoline cars. They were nothing like a 2CV. So, it seemed that the problem with traffic air pollution was not due to the cars and vans, but to the poor and outdated test. It was the government's fault, or so we were led to believe. Then along came the VW scandal.

Huge pressure was piled on governments to find out if any other car makers were cheating. The UK's first response was to write to all of them and ask. In response to public pressure, European governments began to test diesel cars themselves rather than rely on the manufacturers' tests. A chance finding in some of the investigations finally shed light on the mismatch between our city air and what had been expected in policy.

This was clearest in the UK tests.[14] First, cars were tested on the official cycle in a laboratory. They all passed as expected. The analysts then varied the test a little to trick any recognition software. They started the test with the fast driving section rather than the slower urban simulation. A vehicle made by the VW group (actually a Škoda Octavia) did not recognize this as a standard test and emitted so much nitrogen oxide that it failed. Cars from other manufacturers and a newer VW group car passed the modified test, so it looked as if the problem was confined to older VWs. Next the tests were done in the normal way, except with a warm rather than a cold engine. Oddly the average emissions were greater when an engine was cold, some cars producing over 2.4 times more. Finally, the cars were taken outside and driven under the same conditions as simulated in the tests. Emissions were on average four to five times higher than in the official indoor test. It destroyed the idea that the mismatch between the tests was solely because the laboratory cycle was too gentle and undemanding for modern cars. There had to be another reason why driving a car outside could make such a huge difference to the emissions. After all, cars are made to be used outside on roads and not indoors on rollers.

The UK tests were carried out in winter. If the VW scandal had happened earlier in the year and the tests had taken place in the summer, we might not have discovered that most diesel cars emitted more nitrogen oxides on colder days.[15] Just a few months before the VW scandal, tests in Norway had found inexplicably high levels of nitrogen oxides in car exhausts in Nordic winter conditions.[16] This should have served as a warning, but the results were largely ignored at the time as being from an extreme test. Nearly two years after the

VW scandal, measurements on 9,000 cars in Sweden showed that the nitrogen oxides produced by an average Swedish diesel car at 50° F were around twice those at 77°F.[17] This discrepancy has become known as the "temperature window."

When asked about the outside driving results, manufacturers said that using cleanup technologies in cold temperatures could risk engine damage, preventing the pollution controls from working for the required lifetime of the car. It seems bizarre that vehicles could be fitted with exhaust cleanup which was hardly used in order to ensure that the technology worked for the lifetime of the car. Protecting the engine from damage was allowed, so this was perfectly legal, in contrast to the software used by the VW cars to detect if they were being tested.[18] The ICCT experts strongly disagreed with the car makers' claims.[19]

Safety is paramount throughout the vehicle industry, so why did this mandate not extend to the safety of those breathing the exhaust? Why did manufacturers not compete with each other to produce less polluting cars in the same way they competed on crash protection and fuel economy? Point-of-sale information on real-world emissions could help to make pollution a factor in the buyer's choice, but this would have to start with ensuring that real-world emissions matched those in the test.* In 2011, I was part of a working group set up by Environmental Protection UK to devise a labeling scheme for new cars to include their vehicle exhaust. It soon fell flat as car makers cited technical problems with our questions. At the time I did not think any more of this rejection. As the VW

* The UK company Emissions Analytics began publishing test data online in 2017. See www.emissionsanalytics.com.

scandal broke I recalled the stalled scheme and wondered what we would have uncovered.

Tests carried out on diesel cars right after the VW scandal did reveal some traits suggesting that the future might be better. While the early tightening of the test standards produced little or no improvements in real-world driving, the newest cars that met the 2015 limits (known as Euro 6) emitted on average about half the nitrogen oxides compared with the vehicles that they replaced. However, this was still around seven times more than under test conditions.

One of the consequences of the VW scandal was a new real-world test before a new car type was approved for sale, to be phased in toward 2020. This would catch the worst emitters but still did not mean that new cars would have to match the laboratory test standards when used on the road. The new cars launched in 2017 could still emit more than twice the laboratory test standard.

In 2015, the best cars could already keep within test limits when used on the road. It was technically possible at that time. If regulators had the ambition to ensure new vehicles from 2017 matched the best in class in 2015 then illegal air pollution would be remedied faster. If the current rate of progress between 2010 and 2016 is continued, Parisians will have to wait until around 2035 for their city to meet the 2010 limits and Londoners will have to wait until beyond 2100,[20] ninety years too late.

Some hope for faster improvement in nitrogen dioxide comes from trucks and buses that meet the Euro 6 standard. These standards are different to the ones for cars. On-road tests showed that a fully laden truck could produce far less nitrogen oxides than a Euro

6 car.* This turns our perceptions on its head. It seems incredible that trucks and buses with large engines might emit less pollution than cars, at least with respect to nitrogen oxides. With newer vehicles coming onto the roads, levels of nitrogen dioxide did begin to fall between 2010 and 2016, but policies were not sufficient to bring about improvements in all locations. We need new policies in addition to the natural renewal of vehicle fleets.[21]

The whole family of nitrogen oxides were controlled without also controlling nitrogen dioxide on its own. Rather than going down, the level of exhaust pipe or primary nitrogen dioxide increased from the late 1990s and through the first decade of the twenty-first century.[22] The start of this trend was one of the explanations for the odd data obtained at Marylebone Road in London in 2003.[23] The increase in exhaust pipe nitrogen dioxide was due to technologies such as diesel oxidation catalysts, designed to control carbon monoxide and hydrocarbons in the exhaust, and to methods of preventing particle filters from clogging. An optimal policy for controlling nitrogen dioxide would therefore have controlled not only the total nitrogen oxides but also the nitrogen dioxide in the exhaust pipe.

Before issues with nitrogen dioxide came to the fore, the main concern for diesel exhausts was particle matter. The cleanup technology of choice was the diesel particle filter, which is claimed to be extremely effective. Certainly black carbon or soot particles next to London's roads decreased between 2010 and 2014[24] but, in contrast to the evidence on nitrogen dioxide, there is only limited evidence

* This only works if the truck's exhaust controls are working. Worryingly, in late 2017, UK Department for Transport inspectors found one in thirteen trucks had been fitted with a cheat device in their exhaust system. See https://www.gov.uk/government/news/more-than-100-lorry-operators-caught-deliberately-damaging-air-quality.

that particle filters are working as they should be.[25] We know from Owens' early work and the acid rain issues of the 1970s and 1980s that most of the particles that we breathe form in the air from chemical reactions between pollutants. The nitrogen oxides from Europe's diesel vehicles go on to form one of the major components of our particle exposure, meaning that this is a particle problem too.[26]

Large questions remain about the particles that form in the air from the gaseous pollutants in diesel exhaust. In a later twist in the evidence, a series of experiments in London suggest that currently unregulated hydrocarbons in diesel exhaust could be making an important addition to the particles that are forming in urban air in Europe.[27]

So, where next for the diesel vehicle? The post-VW erosion of public confidence in diesel vehicles, the difficulties of controlling their exhausts and ever tighter standards might bring about cleaner diesels. The extra cost and space required to fit abatement technologies could see diesel retreat toward larger car sizes, leaving smaller cars to run on cleaner hybrid-gasoline engines, following the Japanese route. In 2018, there were signs that new car buyers were shifting away from diesel, but Europe is unlikely to abandon diesel cars until government subsidies are withdrawn. In the wake of the VW scandal, Paris and several other cities committed themselves to phasing out diesel by 2030. It is difficult to see how this can be achieved given the lack of current alternatives to diesel for buses, trucks and other heavy vehicles, but hopefully this city pledge can drive innovation. In 2017, the UK government committed to an end to new gasoline and diesel cars by 2040, but interestingly this commitment does not include heavy vehicles.

The problem with diesel cars is especially serious in Europe. Speaking in London in 2004, one of the authors of the Six Cities study predicted that Europe would regret its experiment with the diesel car. Diesel failed to deliver on the promised climate benefits, and air pollution from diesels will have led to many hundreds of thousands of early deaths since the 1990s. Even commercially we should question the success of the diesel car. Although diesels sold well at home due to tax advantages, European car manufacturers struggled to sell their diesel cars in US and Japanese markets. Going back to the start of the diesel story, there is perhaps a silver lining. The dieselization of Europe's car fleet came about because natural gas replaced oil in heating and power generation. Compared with burning oil, this increased use of natural gas will have reduced climate change impacts and harmful air pollution from buildings and industry, but was this advantage worth the health and climate impacts of converting the heating oil to diesel and using it in our cars?

There are questions for policymakers too. Why did they not act earlier in the face of evidence that policies to control air pollution from traffic were not working? Instead, they kept faith with the vehicle manufacturers, and time after time they simply relied on the next, tighter stage of Euro exhaust standards to save the day when the last ones had clearly failed. It also seems astonishing that governments continued to trust the vehicle industry's assurances when all the evidence told a different story. Once again, the voice of the polluter was allowed to drown out the voices of the environmental and health scientists. Manufacturers argue that they need long-term pathways to develop technologies and bring new cars to market, but similarly, they should also take responsibility for ensuring that the products they sell do as little harm to health as possible. Given the tax breaks applied to diesels, car makers owe us all an explanation as to why their products met the legal test standards yet were actually spewing out unacceptable levels of pollution when driven past our homes and schools.

Chapter 11

Wood-burning—the most natural way to heat my home?

Beneath the radar, an old problem returned to Europe in the twenty-first century. Gone are home fireplaces fueled by coal. Instead, browse through the home-style magazines at your newsstand or watch interior design programs on television and you will see beautifully decorated living rooms complete with a roaring wood fire. This must-have design feature comes at an environmental cost.

The rediscovery of wood-burning in the cities of northwest Europe can be traced to Paris. In 2005, Olivier Favez was a young student researching for his PhD when he struck upon something unusual and worrying.[1] He was measuring air pollution on the edge of Paris's Parc de Choisy.* Typical of the small parks and green spaces that add to Paris's character, the Parc de Choisy has formal

* Around two miles from the location of the Observatoire de Montsouris, where the world's first long-term ozone measurements were made in the late 1800s. See Chapter 5.

paths and lines of trees, fountains and a playground. On one side is a large brick building that houses the city's public health laboratories, from whose roof Favez was taking measurements. Unsurprisingly his instruments told him that the air contained lots of soot from diesel traffic. But Favez noticed another pattern, one that had been seen before but that he did not expect to find in the center of the French capital. The instrument had been used previously in Alpine valleys, where wood-burning has a serious impact on air pollution, so Favez recognized the signal right away. But if it was correct then somehow wood-burning was a serious problem in Paris too. Favez continued his measurements for five weeks and each night, especially on weekends, he saw air pollution from wood-burning. It looked as if wood smoke was adding an extra 10 to 20 percent to the city's particle pollution. Not only this, but the wood-burning pollution was coming from within the city itself, not drifting in from the countryside.

Gradually traces of wood-burning were found in the air in other major cities in Western Europe, often when scientists were looking for something else. In 2010, I was working in Paris as part of an international advisory panel on the city's air pollution. The panel was led by Martin Lutz, who had formerly worked in the European Commission and was now heading the air pollution section of the Berlin city government. We were looking carefully at Favez's data and new measurements that were also showing a widespread wood-burning problem. We all began to think of our home cities and how widespread the issue could be. In one memorable meeting, Lutz declared confidently that wood-burning was not a problem in his city. He underestimated the new popularity of home fires.

Unknown to all of us, back in Berlin, PhD student Sandra Wagener was investigating how the city's trees and plants affect its air pollution.[2] She collected particle samples from the air in three different parts of the city and took them back to her laboratory for analysis. One of the chemicals that she looked for was a sugar called levoglucosan. In much the same way that caramelizing onions makes them taste sweet, levoglucosan is released when wood is burnt. Wagener found plenty of it, and not just in the leafy suburbs; wood-burning was spread throughout the city.

Gradually, across Germany, France and Belgium, the same story was repeated time after time. City governments had assumed that wood-burning was a thing of the past but scientists measuring the air that people breathed were proving them wrong.

London's story is a little different. In 2008, with climate change coming to the fore, the UK government created legally binding targets for carbon emissions. It was rapidly realized that an 80 percent reduction in greenhouse gas emissions by 2050 could not be achieved without changing how we heat our homes, schools and offices. Heating with renewable electricity was one solution but this would require huge increases in renewable power. Another solution was renewable heating* using solar panels, heat pumps and wood-burning, supported by government subsidies. Ideally the wood would be burnt in high-efficiency modern boilers with cleanup systems rather than in stoves and fires at home,[3] but whatever path was taken, I was certain that London's air was going to be changing again in the next decade. I thought it would be a good idea to collect some baseline data against which these changes could be judged.

* And the so-called Merton Rule and the Renewable Heat Incentive was born.

In winter 2010, my team at King's College London placed samplers along a twenty-two-mile line across London, running from Ealing in the west to Bexley in the east. Timothy Baker installed samplers and Anja Tremper prepared the filters. Some days were very cold, and we took it in turns to go out and collect the filters, standing on top of ladders as the heat was sapped from our hands. Each filter was taken back to the laboratory, carefully wrapped in foil and frozen. When the experiment ended, the samples were carefully packed in a giant cool box and flown to Norway for analysis.

We had to wait two months for the Norwegian laboratory to do their work. I spent many sleepless nights worrying if the experiment had worked; a lot of money had been spent, but was there enough wood-burning to detect? I could see myself standing in front of the funders and declaring, with a red face and a shrug of my shoulders, that there might be *some* wood-burning pollution in London's air, but I could not say how much.

When the email finally came, it contained a big surprise. In the now familiar pattern, it was clear that wood-burning was already strongly prevalent in London. I set to work with some calculations. It turned out that wood-burning, a source that did not officially exist, was making up 10 percent of the particle pollution that Londoners were breathing during winter.[4] There was some other information too. Wood-burning largely happened on weekends. Londoners were not burning wood to heat their homes every evening; they were mainly using wood as a decorative or extra heating source.

London's low emission zone had come into force two years earlier. It was a big step in improving the city's air, so how did the wood-burning compare? I set to work with some more calculations.

It turned out that the extra particle pollution from wood-burning not only reversed the predicted pollution gains from the first two phases of London's low emission zone but was six times more than the particle pollution that we had saved.

Wood-burning was important, and something needed to be done—otherwise the money that we were investing in cleaning up transportation and industry would be negated by people burning wood at home. With new government inducements to burn wood, the situation would only get worse. I set out to tell people. I presented my data to the environment ministry and at meetings and conferences around the UK and Europe. I teamed up with scientists from Berlin and Paris and we published a warning article called "Time to tackle urban wood-burning,"[5] but local, city and central government were focused on traffic pollution and no one wanted to hear that they had another problem.

Then, in 2015, a UK government survey revealed that around one in twelve UK homes was burning wood. Suddenly official pollution emissions for the UK were revised to say that wood-burning was producing 2.6 times more particle emissions than traffic exhaust[6] and wood-burning was finally recognized as a problem.

At King's College London, we have a very large database of UK air pollution measurements. Some of these (about 52 million) came from the same type of instrument that Favez had used in Paris in 2005. However, we had mainly used the results to look at black carbon from diesel traffic. It struck me that we could do the same thing as Favez and work out how much pollution from wood-burning was in the air across the UK, going back over nearly eight years to when measurement began in 2009. We could get results quickly rather than waiting to collect samples and sending them off to Norway. Even better, I would not have to brave the cold weather climbing ladders to collect samples.

My colleague Anna Font and I set to work.[7] Our data center has some powerful computers and many millions of calculations later we got our result. Wood-burning was adding between 3 and 17 percent to winter particle pollution in cities throughout the UK mainland. Oddly, despite sales of nearly 1.5 million wood stoves in the UK, particle pollution from wood-burning was not increasing; it was stable or even going slightly downward. How could this be? One possible explanation was that wood-burning had restarted earlier than we thought, with people using the old fireplaces that were still present in most homes. By the time we began measurements in 2009, new wood stoves were starting to replace wood-burning in open fires. A modern stove produces less than a quarter of the pollution compared with an open fire.

Our flat trend might have been due to two factors that balanced each other out. People who had begun burning wood in open fireplaces upgraded to stoves, reducing the wood smoke problem; but this was offset by more people joining the fashion for home wood-burning. It was clear that habits had changed too. In 2009 and 2010, wood-burning happened mainly on weekends. By 2016, it was every evening. This fit the idea that fireplaces might be used for family time on the weekends, but the 1.5 million householders who'd installed new stoves would be more likely to use them every night to justify their investment.

An additional problem with wood-burning is where and when it takes place. People burn wood in their neighborhoods at times when everyone else is at home. As wood smoke builds up in an area, it drifts into everyone's houses and lots of people are exposed. A study in Vancouver found that even modest wood-burning in residential areas can lead to greater air pollution exposure than that from busy roads, where most people only spend a short time.[8]

As an air pollution researcher, I've noticed a change in the emails that I get from members of the public. These used to focus on traffic but now neighborhood wood-burning dominates. Typically, the emails come from carers, people who look after an elderly relative at home or parents. There are cases of children's bedrooms being filled by wood smoke from their neighbor's chimney. I am sure that these are just the tip of the iceberg.

The biggest UK cities still have the smoke control laws that were put in place following the 1952 London smog. These also ban wood-burning in open fires. Almost all of London is a smoke control area, but in 2015, 68 percent of wood-burning homes in London were using an open fire. Clearly the law has fallen into abeyance. In many ways, controlling wood-burning should be easier than it was to control coal-burning in the 1950s and 1960s. At this time people had little option other than solid fuel fires to heat their homes, but since the 1970s, most UK homes have gas or electric heating. There is no need to burn wood in our cities.

Ten years after Favez's discovery of substantial wood-burning in Paris, the city got within a hair's breadth of a ban on wood-burning in open fires. With just days to go before a ban was put in place at the start of 2015, the French ecology minister Ségolène Royal attacked it as "ridiculous" in a serious of extraordinary statements, even though the policy had originated in her own department.[9]

Although wood-burning produced more particle pollution than traffic exhaust, banning it was deemed an overreaction; banning a romantic evening with a glass of wine in front of the fire was portrayed as an attack on the French way of life. Once again we run into the nostalgic idea of the hearth being at the heart of a home and the politically unpalatable territory of depriving a homeowner of the joy of their fire. This one-sided narrative delayed the cleanup of the smogs in the 1950s and the same pattern is being repeated. It ignores the use of our air as a waste disposal route and its impact on the neighborhood. Clearly, air pollution scientists have a long way to go before enough politicians and the public can be convinced of the impacts from home wood-burning.

So why do people burn wood? This is obvious to anyone who has sat in front of a solid fuel fire, felt the warmth on their cheeks and toes and watched the gently dancing flames. It is relaxing, cozy and somehow wholesome and reassuring. When Danes were interviewed about why they used wood-burning stoves, they listed comfort, coziness and relaxation as the first reasons.[10] District heating or heating with electric or gas is available at the flick of a switch or the turn of a thermostat dial, but wood-burning connects us with the process of making our homes comfortable. Chopping and seasoning wood can be a family activity, and Danes liked to heat their own homes independently. In much the same way as we like to cook at home rather than buy ready-made meals from the supermarket, burning wood is seen as work to make our lives better.

With much in the media about climate change and the pressure to conserve energy, the carbon neutrality of wood was the third reason given. Most Danes thought that wood smoke was less harmful than other pollution and they associated it with happy childhood memories. Yes, Danes also noticed bad smells from wood-burning, but it was always from a neighbor's fire, not their own. Similar reasons were given by Australians in New South Wales, who listed the comfort and the satisfaction of heating a home themselves as their top reasons. Here people were very aware of the annoyance of wood smoke but, again, this came from other people's houses and from outsiders, those new to the town or living in rented homes. The pollution that filled their towns was thought of as natural fog and not smog.

In Australia, an additional dimension was a connection to a bygone rural way of life through the woodman who sold logs. A survey in the wine-growing Hunter Valley found that around 30 percent of people who burnt wood knew that it was harming their neighbors' health, but only 18 percent of wood burners were prepared to listen and were likely to change their behavior—and only if it was not too onerous.[11] Despite the estimated annual health cost of AU$8 billion from wood-burning pollution in New South Wales, people were confused by messages from the government and in the media. On the one hand they were told about the benefits of the carbon neutrality of wood heating but on the other hand they were warned about the health harm. Faced with this contradictory advice they stuck with doing what they liked to do, which was burning wood.

When we think of New Zealand we think of a pristine landscape of greenery, mountains, rivers and fjords that produces excellent wines, quality food, opportunities for bungee jumping, film sets for *The Lord of the Rings* and very good rugby players. The reality does not quite match the marketed image. The country struggles with the impact of farming and agriculture on the quality of its environment, including its streams and rivers.[12]

New Zealand's air is good by world standards. This is helped a lot by the country's distance from anywhere else. Unlike most of Asia, Europe and North America, New Zealand does not have to contend with large pollution influxes from its near neighbors. Its cities have lots of traffic, but they are not clogged with diesel cars that confound efforts to clean the air of European cities. With limited availability of natural gas and expensive electricity, many New Zealanders, especially those in the South Island, rely on wood-burning to heat their homes. The 2013 census found that 546,000 homes (36 percent) used wood for heating, a similar number to Norway and Denmark; these fires could burn more than 13,000 tons of wood a day during winter.[13] As a consequence, particle pollution builds up on still winter nights.

In the winter of 2016, I was lucky enough to be invited to New Zealand as part of an international delegation of wood-burning researchers.* We toured the country much like a regional theater group, presenting our work. We spoke to the national government in Wellington, to the regional government in Canterbury and even did a lunchtime gig in a community hall in the former gold rush town of Arrowtown in the mountainous south. It was great to hear a range of views and see the problem from different perspectives. The presentations and discussions in Arrowtown were our last meeting. We took a few hours off and headed south to Clyde and Alexandria, where we walked up the hills overlooking the town to watch the sunset. It was beautiful—but as the sun went down, in the cold night

* Huge thanks to NIWA, and especially to Guy Coulson and Ian Longley.

air below us, one by one, columns of smoke began to rise from each house. Slowly the town filled with smoke. Hemmed in by the hills, it had no escape.

New Zealand's national standards for particle pollution allow for one polluted day per year, but Christchurch measured five in 2016 and the city of Timaru breached standards on twenty-seven days. In some places, wood heating makes up 90 percent of wintertime particle pollution. Heating is not optional. Many New Zealanders wake in the morning to see their own breath indoors, and the country's poorly insulated homes and fuel poverty contribute to high winter death rates and juvenile asthma. Double glazing and central heating are not the norm as they are in Europe or the United States, and many householders heat only one room, with a log burner or perhaps an electric heater. The impacts were clear. We visited homes around Christchurch and talked to people whose houses were filled each night by their neighbors' wood smoke. After one visit, we walked past the local school and saw children practicing netball as the smoke from wood fires drifted across the playing court and slowly enveloped them.

The problem is not under-researched. Christchurch has been a focus for investigations of wood-burning for decades. New stoves must meet tight emissions standards and wood-burning pollution is often debated, but it still persists. The reasons why New Zealanders burn wood include the same comfort and coziness motivations found elsewhere, but two further factors resonate here. The first is resilience and being off grid when power supplies are disrupted by natural events such as bad weather or earthquakes. The second is articulated neatly in a paper by Julie Cupples and colleagues at the University of Canterbury titled, "'Put on a jacket, you wuss': Cultural identities, home heating, and air pollution in Christchurch, New Zealand."[14] Cupples found that the same male-dominated and pioneering mentality that leads the "Kiwi bloke" to wear shorts in winter and never to use an umbrella permeated attitudes on home heating, favoring free wood over expensive electricity and regarding investment in insulation as an unnecessary luxury.

Like the U-turn on the Parisian ban on wood-burning, this situation highlights the need to make the case to the public about the harm caused from wood-burning. Knowing about the problem, people should act rationally and abandon their bad ways. But we know that this does not happen. People still smoke cigarettes, drive too fast in their cars and eat sugary foods despite knowing that it is bad for them. We need another approach.

Proponents of "nudge" theory think differently.* Based on ideas of behavioral science, politics and economics, "nudge" uses positive reinforcement or indirect suggestions to try to bring about change, in much the same way that marketing people sell us cigarettes, fast cars and sugary foods. You experience "nudge" each time you go into the supermarket for milk and bread and come out with sweets and chocolates. The regional council for the Christchurch area in New Zealand have been trying "nudge" ideas to help reduce wood-burning pollution. Examples include challenging people to improve their fire-lighting and to show that they have the best wood-burning techniques in their neighborhood.† Videos and

* "Nudge" was born from the contradiction between a libertarian view of government that seeks minimal interference in people's lives and a paternalistic view that people need to be guided toward behaviors to make society better. It won Richard H. Thaler a Nobel Prize in 2017.

† See https://www.warmercheaper.co.nz/.

classes were provided to encourage people to try new ideas, such as lighting their fires from the top of the stacked wood and using plenty of kindling. A lack of kindling is one of the reasons why wood burners sometimes smoke when they are first lit, as we saw so clearly over the New Zealand towns we visited. The provision of free kindling is one answer being tried in Christchurch. But the main problem is that chopping wood for kindling takes time and often leads to chopped fingers.

The need for kindling spawned a simple innovation in 2011. Kiwis are a resourceful bunch, and having seen her mum slice a finger while cutting wood for kindling, thirteen-year-old Ayla Hutchinson had a brilliant idea. Rather than risk axe and fingers, she separated the two and upended the whole process. Hutchinson took an axe head, welded it to a base plate, placed it on the ground and surrounded it with a guide frame. Wood could be hammered down onto the axe blade with a blunt mallet, splitting it. The Kindling Cracker was born. In five short years, Hutchinson's Kindling Cracker went from school science project to a production run of 10,000 per month.

It remains to be seen how much these ideas will help, but improving wood-burning techniques is just one arm of the work being done in Christchurch. This is coupled with increasingly strict emission standards on new wood stoves, rather than an outright ban.

Cleaner stoves are seen as one way to reduce wood-burning pollution in many parts of the developed world. Open fires are the most polluting way to burn wood. They produce roughly two to four times more particulate pollution than even the oldest stove designs. The most modern stoves and wood pellet burners are better still and can produce less than one-fifth of the particle pollution that comes from an open fire. So, upgrading from fireplaces to stoves and from old stoves to modern ones that pass pollution tests should reduce air pollution from wood-burning.

There is one major problem with this approach and that is the long lifetime of fireplaces and stoves. People hardly ever buy a new one. The 68 percent of people who burn wood in open fires in London will have used the fireplace that was built with their house decades or even a century ago. Stoves can be used for just as long, so simply setting standards for new ones will commit a town or city to polluted air for decades to come, unless restrictions are placed on using existing fireplaces and stoves that do not meet the new standard.

Politicians are often reluctant to tell people what they can and can't do in their own homes. One notable exception to this is Montreal, Canada, where ice storms in 1998 left the city without power and people turned to burning wood as a fallback. Installing wood heating as a contingency, and an associated fad for decorative fireplaces, took wood smoke to 39 percent of the city's particle pollution in just a few short years. The city's air pollution was simply out of control, a situation similar to that in the UK in the first and second decades of the twentieth century. The response was a ban in 2018 on all but the most modern wood stoves.[15]

Another approach is stove swap-out or scrappage, funded by government. In scrappage schemes, money is given to householders to upgrade their stoves or fireplaces. The first community-wide stove scrappage scheme took place in Crested Butte, a former gold rush town in Colorado, United States. The town is now famous for skiing and mountain biking but in air pollution circles it holds the distinction of having achieved a 60 percent reduction in wintertime air pollution. This came from a scheme in 1989 to 1990 where almost half of the town's old wood burners were replaced by newer units, and another third were either removed or disabled. Other stove swap-out schemes have taken place in Seattle and Reno.

Perhaps the most studied is the scheme in Libby, Montana. In many ways, Libby is typical of the small communities in the northern Rocky Mountains. Here the majority of people rely on wood to heat their homes. The local geography has been described as a bathtub, with Libby located at the bottom. The early settlers chose the location as it was sheltered from the biting winter winds, but a downside was the air pollution that built up each winter in the valley.

Libby's story is one of the best of times followed by the worst of times. During the 1920s and 1930s, Libby had been a thriving gold, silver and lead mining town, but as these resources dwindled, the town's wealth was sustained by its vermiculite mine. Vermiculite is mined as a rock, but is heated to turn it into flakes that resemble fish scales and can be used as insulation, in cement boards or as a soil improver. At one point Libby's mine produced 80 percent of the world's supply of this mineral. When the mine closed in 1990, the town's economy took a big hit; with the closure of the saw mill in 2002, all its large employers had gone, leaving many in poverty.

To compound the town's problems, the locally mined vermiculite was later found to contain a form of asbestos. Initially this was not recognized as having the same harmful effects as other types of asbestos, but soon after the mine closed mesothelioma, an obscure form of asbestos-related cancer, began to appear in the local population. Free vermiculite from the factory had been used throughout the town: in gardens, driveways, on the baseball field and school running track and as insulation in homes.[16] The impact on the town's health soon extended beyond the mine workers to their partners and children.

Before the stove swap, 82 percent of the town's particle pollution had come from wood-burning. On top of this, many families were struggling to afford enough wood to heat their homes each winter with their old, inefficient stoves. A major investment program was set up to replace all of the town's wood stoves at a cost of over $2.5 million. This would clean the air and also provide more efficient wood burners to help struggling families afford fuel.

Over 1,100 stoves were replaced, rebuilt or surrendered. The majority were replaced with wood stoves and pellet boilers that met the modern standards and around 8 percent of people moved away from wood-burning altogether. Wintertime particle pollution decreased by 27 percent, allowing the town to meet US legal limits.[17]

Tony Ward and colleagues from the University of Montana investigated the impact on the town's children, all born after the mine closure. After the stove swap-out, the town's children experienced less wheeze and fewer respiratory infections and sore throats.[18] Interestingly, the benefits were not only seen in the families that had wood stoves at home; the improvement in air pollution applied throughout the town. This showed the way wood smoke can affect whole neighborhoods.

So new, less polluting stoves can help if the scheme is big enough, but a downside of stove swap-outs is that the new stoves are still not pollution free. The wood smoke in Libby did not go away completely. Stoves are getting better: newer stoves are designed with air intakes at different points to make sure that the wood is burnt properly. Stoves sold in Europe will, by 2022, have to meet Ecodesign standards that set limits on how much smoke a stove can produce. However, as with diesel vehicles, there is a very large disparity between test performance and the smoke that comes from stoves in real-world use. Like car tests, those on stoves are performed in very idealized conditions, using dry wood burnt for just an hour or so rather than the variety of wood that people use at home with frequent refueling and adjustment to keep a fire going all evening.

Most of the data on real-world wood-burning comes from Guy Coulson and his team from New Zealand.[19] Coulson grew up in the UK and worked as a researcher for the British Antarctic Survey and at the University of Essex. In 2005, he upped stakes, relocated to New Zealand and met a new set of air quality challenges. Between 2005 and 2009, his team climbed onto the roofs of over fifty New Zealand homes to measure what came out of their chimneys. The team traveled around both the North and the South Island. A big blue trunk of monitoring equipment was placed in each household's garden and pipework was connected to their chimney. A diary and a set of scales was given to each homeowner to place under their wood basket, and samples of their wood were taken for analysis.

The results were nothing like those from the laboratory tests. On average there was nearly ten times more wood smoke. There was huge variability too, even from the same stove. Some days the emissions were close to those of the laboratory test and at other times they would be up to sixteen times higher. It was a big puzzle to find out why. No single cause could be found, but using wet wood was thought to increase the pollution, as did closing the air vents on the stove. The biggest factor, though, was the person who lit the fire. This was one of the driving factors behind the public information and education campaigns in Christchurch and across the Canterbury region, but one important lesson from Coulson's work is the limitation of what can be achieved by tighter standards for stoves and by stove swap-outs; even modern stoves can produce a lot of pollution.

Burn bans are another way to control air pollution from wood-burning. The idea here is that people are allowed to burn wood on windy days when the air pollution blows away but wood-burning is banned on days when air pollution can build up. Burn bans have been used in Washington State, in the Puget Sound and in Seattle, since the late 1980s. These use a two-stage approach. In stage one, only stoves that meet Environmental Protection Agency standards can be used and in stage two, no wood-burning is allowed unless it is the only way that a homeowner can heat their home. As you would expect, enforcement is key to making sure that these bans work. It is very hard to see with the naked eye if someone is burning wood on a cold winter's night. Coming into homes to check would be time-consuming and intrusive, so instead, enforcement officers tour the city with thermal imaging cameras looking for hot chimneys. Fines of $1,000 are issued.

Burn bans are also used in California's San Joaquin Valley. California is not all sun-kissed beaches; the San Joaquin Valley has dry hot summers but rainy, cold, foggy winters. Census results from 2000 and 2010 showed that just under 10 percent of homes used wood heating, but these homes were responsible for over 80 percent of the region's particle emissions in winter. At their peak, twenty-three tons of particle pollution came from home fires each day. Burn bans had been used in the valley since the 1990s but without much effect. It took a tightening of the bans to make a real difference. In 2003, burn bans increased from fifteen per year to around a hundred, covering most of winter. The Check Before You Burn website now tells people whether they can light their fires or not. Particle pollution dropped by between 11 and 15 percent and the number of days when air quality was poor fell from 35 to 12 percent in response to the increased bans. And, importantly, the number of older people admitted to the hospital with different types of heart problems dropped by between 7 and 11 percent.[20] It looks as if properly enforced burn bans can work, and more positively, the San Joaquin Valley bans also encouraged people to stop burning wood on other days too.

Another good example comes from Australia's island state of Tasmania. Fay Johnston came to Tasmania from Australia's climatic opposite in the Northern Territory, where she had worked with rural communities for twenty years. Writing in *The Conversation*, Johnston talked about the home that her family bought when she moved to Tasmania: "We found a lovely older style house, with high ceilings and verandas, that was heated with wood. The fire warmed the entire house and I was delighted with it. But there was a problem." Wood-burning had become popular across Tasmania during the late 1980s and early 1990s. This caused winter-time pollution problems, especially in Launceston, which is located in a river valley that limits the dispersion of air pollution. The state government realized that something needed to be done to get the city's air pollution back under control and that the air pollution from wood-burning had to be cut in half. Rather than focus on improving stove standards, the solution was to encourage people to switch to electric heating. Advertisements began to appear with the headline "Isn't it time you gave up smoking?" and grants of AU$500 were offered to homeowners.

It was very successful. The number of homes heated with wood was reduced from 66 percent to 30 percent and particle pollution in the air dropped by 40 percent. Johnston and her colleagues found that wood-burning had been having a big effect on health.[21] With wood smoke almost halved, winter death rates dropped by around 11 percent, the change being clearest among men. The improvements were seen both when it came to cardiac deaths and respiratory problems.

As with all these studies, it can be hard to link cause and effect. In this case Johnston compared her Launceston data with that from Hobart, just 125 miles away. Hobart did not have a scheme to encourage people to give up wood-burning, and there no health improvements were seen. The work in Launceston did not finish there. Later schemes focused on improving the way that people burnt wood rather than encouraging them to swap for electricity. Sadly, this had no measurable effect.[22]

What you burn matters a lot. Worrying evidence comes from measurements made at a bowling club in a small New Zealand town. For just over two years, Perry Davy, Bill Trompetter and colleagues investigated air pollution in Wainuiomata, a town of around 16,000 people near Wellington.[23] Lawn bowls plays an important part in the life of many New Zealand towns, with some clubs dating back over a hundred years. At one end of the Wainuiomata bowling green, close to an attractive flower bed, is a shelter for players to keep dry in the rain or stay out of the sun. Behind this sits a large, modern-looking white container, its roof festooned with sampling equipment and a device that looks like Robby the Robot from the 1950s film *Forbidden Planet*. This measured the town's air pollution. As expected, the town's air was full of wood smoke throughout the winter, but when Davy and Trompetter looked at the chemicals in the smoke, they had a nasty surprise. The smoke contained arsenic particles, in amounts sufficient to break New Zealand regulations and 50 percent greater than the limits set for arsenic in European air. Elsewhere in the world arsenic is sometimes found in the air very close to metal factories and facilities that make batteries, but Wainuiomata was a residential area.

The only possible explanation was that people were burning construction timber, treated with chromated copper arsenate (CCA). CCA was invented in the 1930s to preserve wood from rot and wood-boring insects, which are simply poisoned by the arsenic; when the wood was burnt the arsenic was released into the air. You might have noticed CCA-treated wood at your timber merchant or DIY store. When CCA-treated wood is new it can have a greenish tinge, but as the wood gets older it is very difficult to see if it is treated or not.

New Zealand scientists rapidly found that it was not just a local problem in Wainuiomata. Treated wood was being burnt everywhere. Arsenic is synonymous with murders in Agatha Christie novels; the arsenic in New Zealand's air would not have caused such dramatic demises, but one study estimated that exposure would cause around fifty extra cancer-related deaths across the country's current population.[24]

Arsenic is not the only poisonous metal to be found in the air in wood-burning communities. The global economic crisis that began in 2008 had an extremely severe impact in Greece. The country descended into a profound financial crisis: pensions were cut, taxes were increased and unemployment soared, especially amongst the young. Heating oil was taxed less heavily than diesel, so crooked traders cashed in and began selling heating oil as diesel, pocketing the price difference. To close this loophole the government increased the tax on heating oil. Prices rose by 40 percent and sales of heating oil dropped. An unusually cold winter struck in 2013 and as snow fell on Athens, wood yards sprang up selling illegal logged wood.[25] Particle pollution in Thessaloniki increased by around 30 percent. In Athens the financial crisis initially led to improved air quality as people drove less, but winter wood-burning more than reversed the effect.[26] As in New Zealand, increased arsenic was found in the Athens suburbs, suggesting that people were burning waste construction wood. But there was also an increase in lead particles found in the air as people burned old painted wood and their old furniture to keep warm.

The Greek winter of 2013 was especially hard for people to cope with, but this was not the first time that lead had been found within wood smoke. In the majority of European cities, including those in Italy, Hungary, Germany and Finland, there is extra lead in the air when people are burning wood. Burning waste wood could be more widespread than we think, tarnishing the natural image of wood heating.*

Up to now we have focused on wood-burning in the developed world, but this merely scratches the surface when it comes to the health impacts of burning solid fuels at home. Household wood-burning has a far greater impact in the developing world. Globally it contributed around 2.85 million deaths in 2015. Around 3 billion people in low- and middle-income countries have to rely on wood, straw, dung and other biomass for cooking. For most people this is burnt on a simple fire made of three stones with a cooking pot on top. In low-income countries this household air pollution is the second greatest risk factor for early death. The women who do the majority of the cooking and the elderly who are at home receive the biggest pollution doses, but the impact is also very severe among children. It results in illnesses such as childhood pneumonia that are not seen in the developed world. The worst-affected countries are in Africa and South Asia, where the burning of crop waste often makes things even worse. Unlike the decorative wood-burning in the homes of Western Europe, there are no alternative energy sources in these countries. Instead the solution, for now, must focus on the way that people burn wood, helping them to move from three-stone fires in the center of their homes to cooking stoves and from unvented stoves to stoves with chimneys.

* One alternative explanation comes from Sweden where Peter Molnar and colleagues attached samplers to people in the town of Hagfors as they went about life in a Swedish winter. Here they found lead along with the wood smoke that people breathed, but instead of lead paint, Molnar suggested that the lead might come from the soil that the trees were grown in, reflecting contamination from the lead added for many decades to gasoline. See Molnar, P., Gustafson, P., Johannesson, S., Boman, J., Barregård, L., and Sällsten, G. (2005), "Domestic wood-burning and PM 2.5 trace elements: Personal exposures, indoor and outdoor levels." *Atmospheric Environment*, Vol. 39(14), 2643–53.

Several very large-scale schemes have been tried. China's National Improved Stove Program distributed 180 million stoves with chimneys to people in rural areas and a similar scheme was rolled out to 32 million homes in India. Some schemes have tried to help people move away from cooking with wood altogether. In India, a three-year scheme started in 2016 aims to give fifty million people access to bottled, liquefied petroleum gas. In Ecuador, universal coverage of electricity generated from hydro schemes allowed traditional cooking stoves to be replaced with induction hobs. A study of over 21,000 Chinese farmers found that improved stoves reduced deaths from lung cancer. In fact the benefits from these schemes extend far beyond air pollution: they can also produce long-term economic gains for local communities. If women and children are able to spend less time collecting firewood or preparing dung for cooking, they will have more opportunities to earn money for their families or spend more time on education.[27]

The latest research on wood-burning raises a further concern. We are all familiar with the plumes of smoke that come from the chimney of a house, but what happens to this smoke after a few hours in our air? For the answer we have to look to Switzerland.

Despite its small size, the country has world-leading capabilities in air pollution science and also a lot of air pollution from wood-burning in Alpine valleys. Straddling the River Aare between Basel and Zurich, the Paul Scherrer Institute contains some of Europe's most advanced air pollution laboratories and its best scientists. As Arie Jan Haagen-Smit did when carrying out his early work on the Los Angeles smog, André Prévôt and team investigated air pollution in chambers. These are generally large rooms that contain a massive clear plastic balloon where air pollution can be sealed in and studied.

One day Emily Bruns brought a wood stove into the laboratory for a new experiment. She set fire to some logs and used the smoke to fill one of the chambers. They then waited. Lights were used to simulate the sun and slowly, as the hours passed, they noticed that the wood smoke began to change. The gases and particles in the smoke reacted together to make more pollution particles. In some experiments, the concentration of particle pollution increased by around 60 percent and in yet others it tripled.[28]

Similar results have emerged from studies in Finland. What happens in a sealed chamber might not be the same as in the streets where you live, but if these experiments in any way resemble the real world, then the air pollution from wood-burning is even greater than we imagined.

Looking to the future, as we decarbonize heating systems, wood-burning looks set to replace fossil fuels when it comes to heating our homes and offices. Various energy scenarios have been considered for the UK, but only those with a high proportion of renewable or nuclear energy can allow us to meet our climate change targets. Renewable energy sources such as wind, tidal or solar power involve harvesting energy when it is available and not necessarily when we want to use it. With wood, coal, gas and oil the energy stored in the fuel can be released when it is needed; still, cold nights in winter, for example. Burning wood or biomass is one renewable way to fill the energy gap. My colleagues at King's College London looked at the potential impact of future increases in wood heating in the towns and cities of the UK. Although particle pollution from traffic and industry is expected to decrease, these improvements are likely to be offset by more wood-burning, so that by around 2030, particle pollution in UK cities (where 80 percent of the population live) is expected to be much the same as it was in 2015, bringing to a halt the progress that has been made since the middle of the twentieth century.[29]

But is wood-burning really carbon neutral, or by encouraging it are we falling into the same trap as Europe did with diesel cars? Of course, the burning of wood releases carbon dioxide in the same way as burning fossil fuels does. In fact, for the same amount of heat, wood-burning produces more carbon dioxide than burning coal and around twice as much as natural gas. This is because of the chemical composition of wood and its moisture content. Wood is like a sponge, and even dry wood can consist of around 20 percent water by weight. Unseasoned wood can be 40 percent or more. This water must be driven off in the fire as the wood burns, and that takes energy.

Burning wood releases in just a few minutes the carbon dioxide stored in the trees for decades or centuries. The idea that wood-burning is climate neutral arises from the reabsorption of this carbon dioxide as tree growth. But that process takes time. So, for a period, there is more carbon dioxide in the air from burning wood than if we burnt a fossil fuel. To compare the burning of wood with the burning of fossil fuels we need to think about the two alternative scenarios. In one we burn fossil fuels and the tree is left growing and absorbing carbon dioxide in the forest. In the other, the tree is chopped down and burnt, and another one is planted in its place. In the worst-case scenario, if we cut down and burn mature trees that are at the peak of their carbon-absorbing capacity, the payback time can be more than a century compared to burning fossil fuels and leaving the tree in the forest.* The payback time is fastest if we burn wood offcuts produced by forest management.

But in addition to the carbon dioxide from burning the wood itself, we must add the fossil fuel emissions from forestry machinery and from processing and transporting the wood. The Canadian wood burned in Europe travels around 60 miles by truck, 600 miles by train and 10,000 miles by sea, all powered by fossil fuels. Wood-burning can only therefore be climate neutral in the long term if we plant more trees than we harvest.

The climate change benefits from wood-burning are therefore less clear than they at first seem and we need to be cautious in describing wood-burning as beneficial to the climate. Avoiding the production of additional carbon dioxide from wood-burning over the next few decades could be critical if we are to avoid irreversible climate tipping points and limit the maximum global temperature rise.[30]

Wherever wood is burnt for home heating or cooking we find air pollution problems, and in many cases evidence of health impacts too.[31]† Very often, people learn to simply accept the pollution around them and it becomes invisible. The health implications often become clear only when the wood-burning is removed or reduced.

* This is hotly debated. Advocates of wood energy argue that the tree would have been cut down anyway and used in some other way, e.g. for paper or timber. Those on the other side of the debate argue that wood-burning will increase total wood demand and cause more trees to be felled. If this is the case, then wood-burning will need to be balanced by an increase in forests and the carbon that they store.

† Health impacts from breathing wood smoke have also been found in experiments when people breathe wood smoke in laboratory conditions. Bolling, A.K., et al. (2009), "Health effects of residential wood smoke particles: The importance of combustion conditions and physicochemical particle properties." *Particle and Fibre Toxicology*, Vol 6(29).

There is a great injustice at the heart of the wood smoke problem in Western Europe, where the fashion for recreational or decorative burning is undermining the progress that we have made to clean our air. Like London's air pollution problems of the 1950s the solution will require government to enter the difficult arena of telling us what we can and cannot do in our own homes. However, such action is needed because the smoke that originates in our homes pollutes our neighborhoods. These impacts are the nasty side of the popular image of wood-burning as a cozy thing to do in your own home. In this sense there are parallels to the debate around the ban on tobacco smoking in pubs and restaurants. Like diesel fuel for cars, wood-burning has been marketed as being good for climate change, but even this is being questioned. The winter smogs of the nineteenth and twentieth centuries and the pollution problems evident across the developing world should have taught us about the problems of burning solid fuels in our homes, yet the practice has been allowed to return to the cities of northwest Europe. Particle pollution from burning both wood and coal is hard to control, and the smoke is produced right in the middle of our communities where many people live. The smoke from even small numbers of homes that are burning wood can dominate the particle pollution in a neighborhood or even across a city.

The companies that sell stoves point to new standards and say that wood-burning is fine as long as people use modern stoves. However, even stoves that pass new Ecodesign standards still emit particle pollution. The test limits for Ecodesign stoves still allow the production of around six times more particle pollution than a modern Euro 6 diesel truck, or eighteen times more than a modern diesel car.[32] Is it therefore fair that vehicle owners are paying for exhaust abatement and industry is doing the same for its chimneys, but home wood-burning is effectively offsetting the benefits? It is difficult to justify heating our homes with wood or solid fuel when alternatives exist. Changing attitudes, customs and habits will not be easy, but action needs to be taken.

Chapter 12

The wrong transportation

In the 1950s, car ownership was beyond the reach of most UK and European families. Today, many families in the developed world think of owning a car as essential to provide mobility and access to jobs, shops and leisure. But in the late twentieth century, air pollution in the developed world became synonymous with transportation. City authorities have prioritized the building of more roads to cater to a projected growth in motorized transportation, rather than managing health and environmental issues. This has led to the development of a so-called windscreen view of transportation planning that sees the problem from the car driver's perspective only. It puts car use first, then public transportation, with walking and cycling as afterthoughts.[1] There is no focus on travel alternatives and little attention is paid to the environmental and health impacts.

Within this constrained context, many schemes around the world have sought to tackle air pollution from road transportation, with varying degrees of success.

Ever tighter standards for new vehicles should ensure that they are cleaner than the ones that they replace. This idea is at the heart of efforts to curb traffic pollution around the world. Despite the VW scandal and the failure of European vehicle manufacturers to follow standards to limit nitrogen dioxide, they have been very successful in reducing other pollutants, especially those in gasoline exhaust. The standards originated in Europe or the United States are now used internationally. The Americas shadow US standards and the rest of the world has been mirroring the European pathway—but for one important difference: they are lagging behind by ten years or more. For example, new vehicles sold in most areas of India in 2014 only had to meet the exhaust standards that applied in Europe in 2000, despite the readily available control technologies.

Across Europe, low emission zones have become the preferred tool to accelerate the improvement in urban air pollution. These involve restricting access for older vehicles to a city center or sometimes to a whole city. Low emission or environmental zones started in 1996 when the Swedish cities of Stockholm, Göteberg and Malmö banned old heavy goods vehicles. The first zone outside Sweden was in the Mont Blanc tunnel in 2002 and slowly the idea spread to other countries. By the start of 2015 there were over two hundred such zones in twelve EU countries, including seventy zones in Germany and ninety-two in Italy. Most zones apply to heavy goods vehicles only, although zones in Germany, Greece (Athens) and Portugal also restrict older cars, and Italian zones also apply to motorbikes.[2] The world's largest low emission zone (980 sq. miles) is the London scheme that started in 2008. Paris, Europe's other megacity, joined the low emission zone club in 2017 with a restriction on cars and motorcycles that were made to Euro 1 and 2 emissions standards (pre-2000) and heavy vehicles that did not meet Euro 3 (pre-2006).[3]

It is surprisingly hard to measure if some types of air pollution control work. Many people expect a sudden improvement on day one of a new policy, but that would be unlikely. Simple variations in the weather can make it difficult to see change in such an obvious way and assessing low emission zones poses problems of its own. The first difficulty is that newer vehicles are replacing older ones all the time. The zone only accelerates this trend, and we therefore need to separate this additional benefit from what would have been expected if the zone had not been set up. The second is that the vehicles in an area do not suddenly change on the first day of the zone's operation. In London, fleet operators began upgrading vehicles over six months before the start date. This was great for Londoners, since the benefits came early, but it smoothed the changes, making them harder to detect; the changes on day one of the zone were pretty much zero.[4] The third aspect is the level of ambition in these zones. Businesses and drivers have loud voices in the media and good connections to politicians, so debate about these zones tends to emphasize the costs and detriment to their businesses, curtailing the scale of the schemes. In London, phase three of the low emission zone was delayed for two years due to business objections.

The impacts of zones can also be more complex than envisioned. In 2008, the London zone brought about a decrease in exhaust particles alongside roads in outer London, but in central London no improvement could be found. To understand why, we had to look more closely at the types of traffic in each area. The air pollution from traffic in central London was dominated by buses, all of which had been fitted with particle filters some years before the low emission zone started. The additional changes brought about by the low emission zone in central London were therefore small. By contrast, outer London traffic featured a lot of old heavy goods vehicles and the impact here was much greater.[5] An important lesson to learn is that low emission zones have to be designed around the traffic that already exists in each city. Further small successes followed in London with the third and fourth phases of the London zone in 2012, which resulted in improvements of around 3 percent in particle pollution citywide.[6]

Unlike London, low emission zones in Germany also ban the most polluting cars, with a focus on diesels. These have also been successful. Across the country, differences in air pollution could be detected between zoned and non-zoned cities.[7] This was especially the case for particle pollution. By contrast, researchers struggled to find improvements from Dutch zones. One possible reason is that the schemes in the Netherlands were too weak; they applied only to small areas of each city and banned only the very oldest trucks.[8]

People often fear that banning the most polluting vehicles will cause them to divert around the zone and simply move the problem, rather than cure it. Before London's low emission zone was introduced the capital had one of the oldest delivery fleets in the UK. The zone caused this to change rapidly as operators bought newer vehicles, and thankfully there was no evidence that the old vehicles moved to the other areas around the capital.[9] The same was seen in Germany, where objectors feared that low emissions would make air pollution worse for neighboring areas. It simply did not happen.[10] It appears that low emission zones induce the scrappage of many older vehicles.

If low emission zones are effective, why do European cities have such large problems with nitrogen dioxide? Twenty-one European countries failed to meet the legal requirements for this pollutant when they came into force in 2010. Traffic data shows that the London zone was effective in banning the oldest vehicles.[11] That part of the policy worked. However, as discussed in Chapter 10, the real-world exhausts from new diesel vehicles did not decline in line with the Euro regulations, nullifying the benefit that low emission zones should have brought.[12]

Other types of restriction have focused on reducing the total amount of traffic. Several European cities have schemes that restrict traffic, change parking or subsidize public transportation to control the worst smogs.[13] These schemes are very much like the burn bans used in the western United States when smog is forecast. The largest European scheme is in Paris, and includes speed restrictions on main roads, free or subsidized public transportation and a ban on some cars.* Similar schemes operate in other cities across France and in Belgium. In 2015, Madrid introduced a scheme for emergency actions during smogs and has enacted them frequently. In December 2016, the city had a week of bans on cars according to their number plates, allowing odd-numbered and even-numbered cars to drive on alternate days. In Oslo, diesel cars were banned for two days to combat winter smog in 2016. These temporary bans are controversial and the evidence that they work is thin. From a health

* The Paris scheme also places restrictions on industry and farming depending on the nature of the smog episode.

impact perspective, we know that long-term exposure to air pollution is more harmful than short-term exposure, so it is better to focus on reducing air pollution every day rather than appearing to control it by addressing the worst smogs only.

Unlike the schemes in Europe that are only enacted during smogs, other types of restriction are in place every day and form part of city life for some parts of South America. The first was created in 1986 when Santiago's Restricción Vehicular scheme began banning cars on certain days according to their number plate. This was followed by Mexico City's Hoy No Circula and similar schemes in São Paulo, Brazil and Bogotá, Colombia. Beijing and neighboring Tianjin also implemented driving restrictions during the 2008 Olympics and a modified version of the restriction continued in Beijing after the games. These schemes have had some impact on air pollution, but several factors weaken how well they work. Some schemes only operate at peak times, so that drivers simply shift the time of their journeys. In many cases the restricted area is too small for an air pollution effect to be seen and others have been undermined as people bought additional cars (often older and more polluting ones) to allow them to drive on any day, making the air pollution problems worse.[14]

A good example of another approach is London's congestion charging scheme. This was introduced in 2003 and charges a daily toll on vehicles that enter a small central area of 14 sq. miles, about 1.4 percent of Greater London. The money raised from the charge support improvements in public transportation, mainly bus services, all over the city. Although traffic in the central area decreased by 18 percent in the first year, air pollution changes were hard to detect. Like many traffic restrictions, the small size of the zone and its limited hours of operation constrained its potential to improve the city's air. Another factor was balancing the reduced emissions from having fewer cars with increases from the greater number of buses and taxis that could move more freely in the zone.[15] Look at congestion in central London today and you would be forgiven for assuming that the congestion charge was not working, but what would the city be like if the zone was removed? Certainly there would be less funding for public transportation, but it is hard to say how much the traffic would increase.

An insight into the disastrous effects of taking away a traffic restriction is provided by Jakarta, Indonesia, when high occupancy vehicle lanes were removed in 2017. High occupancy vehicle lanes were first introduced in Washington, New York and California in the 1970s and spread to many cities, both in the United States and internationally. They cannot be used by vehicles with one occupant, and are designed to encourage ride sharing by allowing cars with a passenger to speed past the congestion. Critics say that giving over a whole lane to a small number of vehicles is an inefficient use of road space and that the traffic would flow better without these lanes, but an abrupt end to Jakarta's lanes proved otherwise. The scheme was stopped to prevent a growing trade in the hiring of passengers, especially children, who would wait in lines by the roadsides. When this came to light, the city's politicians had to act fast. Overnight, the scheme was terminated and suddenly the traffic could spread across all lanes. However, instead of improving journey times and congestion, traffic worsened over the whole road network almost immediately. Delays increased by up to three minutes per mile and this effect spread onto roads that had never had high occupancy lanes.[16] Releasing the restriction encouraged more people to use their cars. This is called induced travel.

Induced travel is at the heart of one of the paradoxes related to road building. The phenomenon is familiar to drivers on London's orbital M25 motorway, where the addition of extra lanes seems to do little to ease the traffic flow; instead the enlarged road becomes as congested or even more so than before. This leads to proposals for another lane and so forth. Increasing road capacity is a futile route to curing our road transportation and air pollution problems. There are countless examples of this throughout the UK and further afield.

Two paradoxes are frequently discussed when thinking about road building. The first is Braess' paradox.* This is rooted in game theory and describes the consequences of the addition of a new link in a road network; a bypass around your town, for instance. Drivers each make an individual decision to use the new link in the hope of getting from A to B quicker. While this is sensible for the individual driver, the cumulative effect of everyone making the same decision means that congestion increases on the new route, the overall efficiency of the road network decreases and no one benefits.

The second is Jevons' paradox, which has wide ramifications for environmental economics. In 1865, thirty-year-old William Stanley Jevons,† or just Stanley Jevons as he preferred to be known, was pondering the future of Britain's coal reserves and economic growth.[17]

* Rather than refer to the German original you can read a good summary at https://www.forbes.com/sites/quora/2016/10/20/bad-traffic-blame-braess-paradox/#57129abe14b5.

† Jevons had many interesting thoughts on the challenges of powered flight. He also considered the use of electricity, which he appreciated as a means of transmitting power, and lamented much of the focus on experimentation with static electricity. He also compared natural sources of power such as wind and water, which were subject to the variability of weather, to coal, which could be used at will.

Jevons' family were Liverpool iron merchants but, owing to the collapse of their business, the young Jevons spent his mid-twenties in Australia working in a lucrative job as a chemist at the new national mint. Australia was a new country experiencing a series of gold rushes and there were regular debates over the building of railways. These sparked the young Jevons' interest in economics. Returning to the UK, he could see how the country was being increasingly powered by ever-better steam engines, but what would happen when the coal ran out? The idea that increased efficiency would make coal reserves last longer was popular at the time but roundly dismissed by Jevons, who believed "it is wholly a confusion of ideas to suppose that the economical use of fuel is equivalent to diminished consumption. The very contrary is the truth." More efficient industry would use less coal in the short term, but the reduced cost of goods would drive demand, a rebound effect causing more factories to open and more coal to be consumed.*

The analogy in traffic terms is with travel time. Each time we build a new road or improve travel on an existing route, it reduces the time taken to get from A to B. This makes it more efficient, and therefore attractive, for people to travel between A and B for work and leisure, and for goods to be transported more cheaply. In turn this leads to increased travel demand, in the same way that more efficient use of coal led to more, not less, consumption.

In 1994, a UK government committee came to the radical, but correct, conclusion that building new roads did not ease congestion.[18]

* It is argued that energy or resource efficiency gains need to be used in conjunction with green taxes to prevent cost reductions and produce a societal dividend; or schemes such as cap and trade where a finite number of pollution emissions permits are sold and traded.

It based this finding on detailed studies of many schemes. One of these was London's elevated Westway, which starts in Paddington in west London and carries lanes of traffic overhead, supported on concrete stilts. It was one of the major civil engineering projects of the 1960s and became the UK's longest elevated road. The intention was to carry traffic above people's homes and divert it from the roads below. Instead the Westway was rapidly filled with new traffic, as people made journeys that they could not make before, and the roads beneath remained clogged and congested.

The committee also studied the Blackwall Tunnel in east London. A single tunnel had been built in Blackwall in late Victorian times to carry people and horse-drawn wagons beneath the Thames. It became an important link between Kent and London's East End. In the 1960s a second tunnel bore was built, along with new link roads, to improve the flow of traffic crossing the river. To see if this new tunnel eased congestion, the committee looked at all the traffic crossing the Thames before and after the new tunnel. Congestion on nearby ferries and bridges was not relieved; far from it, rush-hour traffic on river crossings in the east of London increased by 50 percent after the tunnel was opened. This was not due to traffic growth alone; traffic across the bridges in west London grew by less than 10 percent. The new east London tunnel created new journeys.

This effect is not confined to London. Studies on ring roads, including two in northwest England and one in Amsterdam, showed that increased traffic on new roads was not matched by falls across the alternative routes. A 2006 study on three English bypasses found similar results.[19] When traffic in the town center and on the bypasses were added together it was found that the total traffic increased after the new road opened, and town centers remained congested.

The same thing has happened in the United States.[20] In some cases, doubling the number of lanes on a road doubled the traffic. Other new or expanded roads failed to ease congestion because people shifted their journey times. Rather than staggering their journeys to avoid the congested times of day, drivers all headed onto the new roads at the same time and congestion prevailed. It was the Braess paradox in action.

Many new road schemes are justified as support for local or regional economic growth: to move goods in and out of an area more easily, or to improve the attractiveness of an area for business. If the new road capacity is successful in terms of economic benefit, then we would expect an increase in commercial traffic after the road scheme is completed. However, increased commercial traffic is often just a small part of the picture. The main increase in traffic on expanded US roads was personal driving. The number of journeys that people made was not the main reason. People did not make two journeys to work instead of one; instead they decided to use the new road to work or shop further from home.

Can induced travel be made to work in reverse? Is it possible to reduce traffic by removing road capacity and thereby improving air pollution?

In short, yes. A review of seventy schemes in eleven countries showed that traffic reduction resulted from reducing road capacity.[21] These schemes included pedestrianizing the centers of European cities; restrictions on access to the City of London following the IRA bombing in 1993; the closure of bridges such as London's

Westminster Bridge, Tower Bridge and Hammersmith Bridge for repairs; city-center traffic schemes in UK cities; the introduction of bus lanes, including those in the UK and Toronto, Canada; the closure of a rural road in southern England; street enhancement in Norwegian towns; the Tasman Bridge collapse in Hobart, Australia; and the effects of earthquakes in Japan and California, US, where transportation links were suddenly removed from the network. The circumstances were all very different, so it is hard to compare them directly, but in half of the cases, over 11 percent of vehicles disappeared. In some cases the reduction was short-lived and drivers adapted to the new road network, but where road capacity reduction was a permanent part of a wider transportation strategy the reductions were locked in for the longer term. The schemes were not studied for their environmental benefits, but reducing traffic can have multiple rewards beyond reduced air pollution emissions. It can reduce city noise and climate change emissions.

How we move freight around can also affect air pollution. Cities manage and often own the bus, train, streetcar and metro networks that transport people. When it comes to freight, however, it is left to business to deliver their own goods and for private companies to compete for delivery trade. Deliveries often start outside the city, which means that transportation planners view trucks and vans as transient visitors and ignore their needs. Similarly, freight operators view the city as a maze of streets to be navigated as quickly as possible. This leaves freight operators disconnected from the areas in which they work.[22]

A few years ago, I saw this attitude toward freight for myself. I went along to a meeting of transportation planners in central

London. They presented detailed information on how more and more people were working in central London and how they traveled. There was data on nearby tube and bus routes. It was a great analysis. Out of the window, the road outside was clogged with lines of trucks, all queuing at the traffic lights. During questions, I asked the presenter to look out of the window and tell me what all the trucks were doing there. He looked, shrugged his shoulders, and confessed that he didn't have a clue.

The consequences of paying too little attention to freight can be seen on our streets and in the air that we breathe. Vans were the fastest growing vehicle type in the UK between 1996 and 2016, up by a huge 71 percent.[23] Similar patterns were seen across Europe. These new vans were nearly all diesel-powered and shared the same nitrogen dioxide exhaust problems as diesel cars. The growth in van usage is often blamed on internet shopping and home deliveries, but the traffic data says something different. Most of the increase in numbers of vans pre-dated internet shopping; so what are all these vans doing? In 2008, most vans were used to transport equipment, perhaps reflecting changes in self-employment since the 1990s. There have also been massive changes in the way that businesses handle their stocks and supplies. Gone are large central stores of stationery in office basements. Now when I need supplies they are ordered online and a delivery arrives direct from the retailer the following afternoon. This means that our building does not waste space with stores, but "just-in-time" deliveries for industry, retail and offices mean that inefficiencies in storage have been swapped for inefficient transport on our roads. In 2014, 39 percent of vans driving around London were less than one-quarter full.[24] City centers are crisscrossed by mostly empty vans, with competing operators duplicating journeys. Due to intense competition, delivery businesses struggle to comply with employment laws. The structure of the urban delivery market means that large cities have the oldest delivery fleets. An estimated 12,000 light haulage companies operated in central Paris in 2014.[25] Around one in eight deliveries were also made by small businesses directly, especially by shopkeepers using very old vehicles. The free market is not working.

Globally it is hard to find examples of managed freight distribution systems. The central area of Chicago has a dense sixty-mile grid of tunnels that were used until 1959 to collect waste and deliver post, coal and other goods to building basements on a narrow-gauge railway. This became the inspiration for the Post Office railway that operated beneath central London until 2003.[26] In Porto, Portugal, freight streetcars were used to deliver coal and fish and in Dresden, Germany, Volkswagen uses cargo streetcars to move parts from the railway depot to its factory, but examples are few and far between.

Freight consolidation centers are one possible answer. These channel all deliveries to an area in one location. They are then bundled together and the last miles are done with fully loaded, and in some cases electric, vehicles. So far, these centers have not proved economical since the current delivery systems do not bear the full cost of their environmental impact. An efficient national post office monopoly would also help to control the problem. Whatever the solution, there is a clear need for greater planning of urban freight deliveries, in much the same way as cities plan and manage the public transportation of people.

Much excitement has been generated about battery electric vehicles as a cure for our urban air pollution problems. We have used them to move our scientific equipment around London. They are easy and fun to drive. In 2017, the UK and France announced that gasoline and diesel cars would not be sold after 2040.[27] Paris then trumped the national announcements with a proposed 2030 ban.[28] Could this spell the end of the internal combustion engine and herald clean air in our cities? I'm afraid not, or at least not within the next two decades. This is for several reasons. First, and most obviously, we would need renewable or pollution-free electricity to have zero-polluting battery vehicles. Second, the UK ban only applies to cars and not to larger vehicles. There is no plan to mandate a ban on diesel trucks and buses. Third, the exhaust pipe is not the only source of air pollution from road transportation; the particle pollution that comes from the wear on roads, brakes and tires is now a greater problem than exhausts. This is true for all types of vehicles, including electric ones, and particle pollution from wear processes appears to be getting worse. This was shown by my team at King's when we tracked air pollution alongside sixty-five London roads from 2005 to 2015.[29] To our surprise we found some roads where particle pollution was getting worse, not better. These were mainly outer London roads that had increasing numbers of heavy goods vehicles. Here, the air pollution benefits from improvements in diesel exhausts were outweighed by increases in particles that came from the wear on tires, brakes and the road.

The amount of brake, tire and road wear depends on the weight of the vehicle. Accessories such as electric windows and air conditioning mean that new cars can be heavier than the ones that they replace; heavier vehicles cause more road wear and more energy needs to dissipate through their brakes to bring them to a halt.

Brake systems on cars, vans and trucks have also changed. The car that I owned in the 1990s had sealed drum brakes. Thirty years later most cars had open discs. Disc brakes were patented in 1902 but they did not appear in cars for over fifty years, until Jaguar began to experiment with them in their racing cars. Disc brakes were one of the innovations that allowed Jaguar to win the 1953 Le Mans twenty-four-hour race. Because disc brakes stopped their cars in half the distance of the drum brakes that everyone else was using, Jaguar's drivers could wait longer before braking at the corners, overtaking everyone else who had to begin braking earlier.[30] Slowly, reliability issues with disc brakes have been ironed out, and they have replaced drum brakes in the vehicles that we drive on our streets. But this comes with a downside. As the brake discs and pads become hot and begin to wear, they emit tiny particles of metals into the air. This contrasts with drum brakes, in which the wear particles are mostly sealed in.[31]

Toxicologists (including my colleagues Frank Kelly and Ian Mudway) tell us that these tire, road and brake wear particles are far from harmless.[32] When breathed in, the chemical reactions that they cause in our lungs can overwhelm our bodies' natural defenses, leading to lung inflammation and other problems as our immune systems struggle to cope. There are no policies to control these wear particles. Stopping from 30 mph emits around twice the amount of brake particles as stopping from 20 mph, so lower urban speed limits could help, as could reducing traffic volumes.

One study found that the extra weight of the batteries in electric cars means more particle pollution from brake,* road and tire wear compared with similar-sized gasoline or diesel vehicles that we buy today.[33] However, with vehicle range being key to the success and uptake of electric cars, there will be great engineering efforts to make these vehicles lighter (so-called de-weighting design) and therefore their particle emissions will almost certainly decrease in the future. Battery-powered electric cars will move pollution from our cities to distant power plants. This might help air pollution at the street level, but as we saw in Chapter 6, the emissions from power stations can do immense harm. To achieve big benefits we need carbon-free electricity.

We also need to reduce our dependency on road transportation. Around the world, countries are rapidly becoming more urbanized. For these developing countries it is important to ensure that they do not repeat the mistakes of high-income countries. A low-pollution city would be built around walking, cycling and public transportation in much the same way as our forebears built the old cores of many cities before a motorized age. Once a new city is developed around individual motorized transportation, it is very difficult to curb its growth. Examples include the sprawling car-dependent suburbs that surround US cities. Here, low housing density makes it difficult to funnel enough people onto public transportation routes to make them viable, and shops and schools are too far from many homes for people to walk or cycle. Bad city design can therefore lock its residents into a car-dependent lifestyle. There are some good

* Despite regenerative braking in electric vehicles that recharges the car batteries, reducing the use of friction brakes.

examples of cities that have reversed the growth in car dependency, including many towns and cities in Denmark and the Netherlands, which have turned over their car-filled streets and city centers to walking, cycling and public transportation.

Between 1949 and 2012, a massive investment in UK roads and road vehicles led to a tenfold increase in the distance traveled. This has had its downsides. The Royal Commission on Environmental Pollution pointed to a web of environmental and social problems, including air pollution, urban severance for those without car access, a decline in walking and the closure of local shops. The total distance walked each year dropped by 30 percent between 1995 and 2013, and the distance cycled in England and Wales in 2012 was just 20 percent of that in 1952. Importantly, in 2013, 60 percent of car journeys in the UK were under five miles and 40 percent were less than two miles.[34] There is therefore massive potential to convert these journeys to walking, cycling or public transportation.[35]

Writing in *The Lancet*, James Jarrett, James Woodcock and colleagues laid out how active travel could reduce costs to the UK's National Health Service.[36] They imagined a slightly different world. Cars would not be banned, far from it, but we would use them less. They assessed the changes that would happen if the average person in the UK drove six miles per day rather than eight and a half, as they did in 2011. This was a reduction of just two and a half miles. They investigated the potential health benefits if people walked or cycled instead of driving these two and a half miles. This is a walk of about forty-five minutes or a cycle ride of about fifteen minutes. This modest change would lead to savings of £17 billion for the NHS over a twenty-year period, with even greater savings afterward as the population aged. The benefits to quality of life would be huge given the way chronic diseases restrict so many people's wellbeing.

Another view on the benefits from active travel came from the cycle hire scheme in Barcelona. The scheme started in 2007 and by 2009, nearly 40,000 bike trips per day were taken on "Bicing" bikes. We know that cyclists breathe hard while pedaling, which increases their air pollution dose when compared with sitting in a car, and the accident risk is greater. So, would the "Bicing" scheme do more harm than good? Fortunately, Barcelona is home to a global health research center, now housed in a modern building on Barcelona's seafront, giving its occupants probably the best office window view of any scientists in the world. Audrey de Nazelle, Mark Nieuwenhuijsen and their team weighed up the pros and cons of the scheme.[37] The answer was clear. For people who used the bikes the health benefits from the exercise were a massive seventy-seven times bigger than the downsides; the extra risks of accidents and air pollution were tiny compared to the gains from cycling. It was not just the people using the bike scheme who benefited. Everyone else in the city also gained from a reduction in road noise, air pollution and climate change emissions.

Other studies have tried to put a financial value on the benefits of a switch from driving to cycling. If the average European left their car at home and cycled three miles to work (about twenty minutes) instead of driving, the gains to society would be 1,300 euro per year from reduced health costs. This was far greater than the harm from the air pollution that the cyclist breathed and their risk of having a road accident (20 euro per year). The people who shared the same town or city as the newly converted cyclist would be 30 euro per year better off from the reduction in air pollution.[38] The pros and cons of swapping driving for cycling to work or leisure vary according to where you live—the factors include the air pollution and cycling infrastructure—but the benefits from exercise are greater than the air pollution risks in 98 percent of the world's cities.[39]

This is good in theory, but in practice a shift to cycling might be harder to achieve. London's cycle hire scheme started in 2010 and has proved highly successful. Docking stations are located all over central London. The bikes are a far cry from the carbon-fiber racing bikes of the Tour de France but they are very practical. They come complete with lights and a handy front rack for your bag. Pay just £2 and you can pedal between docking stations all day. Londoners took the bikes to their hearts and nicknamed them "Boris Bikes" after Boris Johnson, who was London's mayor when the scheme started.* The original scheme was actually conceived under Johnson's predecessor, Ken Livingstone, and it has been suggested that they should be called "Ken Cycles"—or that under Mayor Sadiq Khan they should be renamed "Sadiq Cycles." The names of their sponsors, Barclays and Santander, seem to have been forgotten.

A health impact assessment was carried out on cycle scheme users between April 2011 and March 2012. During this time, 578,607 individuals made over seven million trips and cycled for over two million hours. Only 6 percent of users had swapped car journeys for cycle hire. Most people took trips that would otherwise have been done by public transportation, by walking or riding their own bicycles. Given that walking is already an active way to travel and

* Boris was a regular cycle commuter during his first term as an MP. I would often see him pedaling along as I cycled near the Houses of Parliament.

public transportation often involves some walking, the benefits of cycle hire might not be as great as the Barcelona study suggested. De Nazelle and Nieuwenhuijsen had assumed that 90 percent of Bicing journeys would have been made by car rather than the 6 percent found in London. However, the increased exercise from cycling still meant that London's cycle scheme was beneficial for public health* and also reduced overcrowding on public transportation.[40]

So, is cycling and daily exercise really that good for us? Between 2006 and 2009, nearly half a million forty- to seventy-year-olds were selected from the National Health Service and interviewed as part of UK-Biobank.[41] The 260,000 people who worked were split into groups according to the way that they commuted. The 16,000 people who cycled as part of their commute were healthiest, with lower rates of heart disease and cancer. They were also living longer compared with those who largely sat in their cars or on public transportation. The 14,000 walkers had lower heart disease than car commuters, reinforcing the benefits of any active travel.

Sadly, it seems that any notion of reducing car travel is off-limits. The fuel protests of 2000 changed the debate about future road transportation in the UK. They began when a group of farmers and haulers from north Wales blockaded the nearby Stanlow oil refinery, objecting to the high cost of road fuel. The cascading action brought the country to a standstill. Protests spread to other distribution centers and panic buying followed. Within four days, 90 percent of the UK's gas stations had run dry and the roads were empty. I recall the peace and tranquility where I live, relieved of the ever-present

* The number of accidents and fatalities was much lower than expected and, despite the heavy design, and the high numbers of occasional users, fatalities were much less than the average for London cyclists.

background noise of road traffic. I met many of my neighbors on the streets during that week and we talked as we all walked to the shops, clearly showing the social benefits of getting out of our cars. As shortages of bread and milk began, the government invoked emergency powers, but the protest ended as swiftly as it began, the organizers declaring that they had made their point.* They had, and their point still resonates nearly twenty years later. A rational discussion about future road transportation has become a no-go area for UK national politics.

Even a 2014 proposal to limit speeds on a section of motorway to reduce air pollution for nearby schools and homes attracted anger from motoring campaigners, who saw it as the start of wider speed restrictions and infringements on the perceived right to drive. Fearful minsters strenuously denied any blanket attempts to reduce speed limits and abandoned the scheme, effectively asserting the motorist's right to pollute over children's rights to healthy air. One motoring organization, the RAC, stated that "a 60mph limit on these stretches of the M1 and M3 is the thin end of the wedge," and another, the AA, said that its abandonment was a "victory for common sense."[42] Even successes are attacked. Policies that included parking restrictions, better bus lanes and car-sharing schemes brought about a 6 percent fall in car ownership over ten years in the UK city of Brighton and Hove, in contrast to a 9 percent increase nationally. This was described as a "disgrace" by the AA, who claimed that it was hitting the poorest and meant that they could not "enjoy the freedom of mobility to shop, or search for jobs, or leave the city limits."[43] The reality is that poorer people often live

* The chronology of the protests is summarized at http://news.bbc.co.uk/1/hi/uk/924574.stm.

on the most polluted roads and therefore gain most from reductions in traffic.

Despite this heated rhetoric, a tide has already turned, quietly and largely unnoticed. In many UK cities, car traffic has decreased. The peak of car use in London was in 1992. Rush-hour traffic entering Birmingham reached a maximum in the mid-1990s and in Manchester the peak was in 2006.[44] This is not just a UK phenomenon. The decoupling of traffic growth from population and economic growth is widespread across the developed world. Observers have coined the term "peak car" to describe the way that car use has reached a maximum and is now in decline. Many reasons have been suggested for this phenomenon. One theory states that our roads are simply saturated with traffic. Others point to wider reasons: more people living in cities, new housing in brownfield areas, better facilities for walking, cycling and public transportation, high fuel costs and the advent of telecommuting. The economic crisis that began in 2008 may be another constraining factor, but peak car predates this.

The change in car use has not been the same everywhere; decline has mainly been seen in cities and city centers rather than in the suburbs and the countryside. The peak-car phenomenon has been largely attributed to young people delaying or shunning car travel. Compared to previous generations they are attaching less importance to car ownership, but will this last? Are the young people at the heart of peak car becoming car-less or merely car-later, delaying getting a car until later in life when they have families? Total UK traffic grew in 2016[45] but there is some evidence of long-term attitude change, with more people in their thirties saying that they do not expect to become car owners.[46] We will have to wait and see.

However, this completely reframes the debate on car use. Rather than the reduction of car traffic being seen as anti-car and a war on personal mobility, instead we can go with the flow. We can focus on those who have shunned car use and those who want to change, thereby normalizing non-car travel lifestyles rather than battling with those engrained in car-focused modes who are resistant to change.

Of course, personal mobility offers huge benefits, including trade, economic growth, jobs, tourism, education and cultural exchange. But we have to be careful to separate mobility from road transportation. The streets of our towns and cities and the roads that connect them will never be free from traffic and they probably never should be. My elderly parents would struggle to live independently without access to road transportation. But it is also clear that many of us are locked into mobility patterns that give us no joy in life. Having to take car journeys for work every day, to get to shops and local services or to take our children to school does not enrich our lives. The drudgery of everyday driving is a far cry from the images of the open road that we are sold by motoring programs and car marketing. These are the journeys that we need to design out of our towns and cities, and our lives.

It is hard to see an area of air pollution control where the direct benefits to public health would be so great as those from a decrease in road travel. Reducing car use and increasing active travel goes beyond changing the way that we use our roads. By encouraging and amplifying emerging trends rather than seeking to start them we have an opportunity to reverse the traffic dominance of our cities to create livable spaces that are not dependent on motorized mobility.

This would be in stark contrast with the self-fulfilling "predict and provide" approach to our road building that plans for ever greater road traffic and then builds roads that induce this change.

A new view of our streets has been embraced by the mayor of London. This sees streets being redesigned to make them attractive public places rather than just through routes for cars and trucks. It envisages our roads, squares and junctions being restored as focal points for communities; places where people want to walk, cycle or just sit on a bench under a tree and talk to friends, and where businesses can thrive. This does not mean excluding vehicles, but it does start with reducing the noise and air pollution from traffic by means including the redesign of pavements and the lowering of speed limits.[47]

For developing countries, the clear lesson is to create livable cities from the outset. This will involve massive challenges for many cities, especially those with millions living in their vast slums—Karachi, Cape Town, Nairobi and Rio de Janeiro, to name but a few. But there is no other choice.

Chapter 13

Cleaning the air

What have we learned from six decades of air pollution management? What solutions have worked and what has failed? There have been some big successes. London is no longer plagued by coal-induced winter smogs and year-long haze. The residents of California do not experience the eye-watering pollution of the 1950s and 1960s. However, new problems have emerged with traffic pollution in the developed world. Here, cities across China, India and East Asia are grappling with air pollution the likes of which they have never experienced before.

It is tempting to think that we could simply clean our air; that the pollution could be filtered or scrubbed out in the places where we live. Many means of doing this have been proposed. In some ways, cleaning the air is similar to trying to take the milk out of your coffee: once it's in, it's in. The main difficulty is simply the volume of the atmosphere. We live at the bottom of an ocean of air, and while our air pollution is mainly emitted close to the ground, it is dispersed in the lower one and a half miles of the atmosphere. That's a lot of cubic miles of air to clean. Our air is also very mobile. The air that we breathe today is not the same as the air that was in our town or city yesterday. There is simply too much of it to deal with.

Undeterred by these practicalities, there have been many proposals to clean the air in specific localities; in busy shopping streets, or around schools where young children are exposed. Plants are put forward as a solution for many of our urban problems: to help areas mitigate and adapt to climate change, to reduce exposure to air pollution and noise, and also to simply make our cities more attractive places to be. In 1661, John Evelyn advocated plantations of shrubs and scented flowers around London to improve the air.[1] Famously, the streets and parks of central London and Paris are full of London plane trees* (*Platanus x acerifolia* or *Platanus x hispanica*) planted in the eighteenth and nineteenth centuries, and the New York Parks Department has the leaf of the London plane as its logo. These trees shed their bark and have shiny leaves, making them resistant to air pollution and blackening by soot. They can therefore survive and look attractive in polluted cities, but can they help to improve our city air?

In 2015, New York celebrated planting one million new trees in just eight years, and Melbourne plans to double its tree cover to 40 percent by 2040. Trees have a very large area of leaves, giving them a total surface area several times greater than the land area that they cover. Planting trees can therefore increase the surface area in a city on which air pollutants are deposited, but a lot of trees are required to make a difference. Typically, urban vegetation reduces air pollution by less than 5 percent.[2] Scientists have tried to work out what

* See Robin Hull's short guide to the London plane tree: http://www.treetree.co.uk/treetree_downloads/The_London_Plane.pdf.

would happen if the most ambitious landscaping schemes possible were implemented: planting all outdoor green space (all the parks and playing fields, along with everyone's garden) with mature trees. Clearly planting this number of trees would be unworkable in a city, but even doing so would only reduce particle concentrations by around 7 to 10 percent.[3] Tests show that hedges and screens alongside roads can reduce pollution concentrations immediately behind them, but despite the planting of so-called green screens between busy roads and children's playgrounds and schools, there is no evidence that the effect extends any useful distance behind the barrier.

If we are not careful, trees and planting can make air pollution worse. Planting trees can shelter city streets from the wind, reducing the dispersion of traffic exhaust and increasing the concentrations for pedestrians and drivers beneath. Trees also produce volatile organic compounds. Pine and eucalyptus trees may have a pleasant, distinctive smell, but the chemicals they excrete create ozone and particle pollution. Other trees and plants contribute too. A study in Berlin showed that the plants could be adding between 5 and 10 percent to the city's ozone. This extra pollution was greatest during heat waves and droughts, when trees produced more ozone-creating chemicals.[4] Yes, we should increase the green space in our cities for many reasons, creating attractive outdoor places to picnic, play, walk and cycle, but we would be deluding ourselves if we relied on trees and plants to clean our city air.

Smog towers are undoubtedly the most conspicuous attempt to clean our air so far. These range from a 7-meter high tower in Rotterdam in the Netherlands to a 100-meter high tower in Xian, China. In the Chinese tower, air is drawn into greenhouses around the base where it is heated by the sun and rises up the tower through filters. The designers claim it to be effective over an area of more than 185 square miles, but the claim founders on a point of physics. The huge-sounding 10 million cubic meters of air that the tower is claimed to clean each day would be less than 0.01 percent of the air of a modest city.[5]

Perhaps the most frequently trialed technique to clean urban air is photocatalytic paints and coatings. In the laboratory these can remove nitrogen oxides and some other pollutant gases when bathed in artificial sunlight. With European cities struggling to reduce nitrogen dioxide from diesel exhausts, simply painting walls and surfaces with photocatalytic paints seems an attractive solution that does not require costly upgrades to millions of vehicles or the political courage to tackle our transportation problems. Painted walls and coated paving slabs have been trialed with various degrees of success. However, no matter how good the chemicals in the coating are, using surfaces to clean the air again flounders on a point of physics: there is a vast volume of polluted air in our cities and it spends very little time in contact with the ground or other surfaces. The UK government's Air Quality Expert Group used a model to test what would happen if photocatalytic paint was applied to every surface in London.[6] Even setting aside shading problems in the low-angle winter sun, the change in nitrogen dioxide would be less than 1 percent and there was concern about the byproducts that would wash off. The surfaces would require regular repainting too. Expending effort to clean already polluted outdoor air will not solve our problem.

Avoiding air pollution can help. Simply traveling along a quiet road or through a park can reduce your personal exposure to traffic pollution by more than half. However, modern air pollution is largely invisible, so it is difficult to know which streets to choose. We can follow our noses, but is it better to walk alongside a busy but open road or along a congested narrow street with high buildings? High-resolution mobile phone maps are already available in Paris, Vancouver and London, but avoiding the sources is impossible when the whole of a city or region is enveloped in particle pollution or ozone. Staying indoors can reduce exposure to outdoor air pollution, especially if you live or work in a modern building with a filtered air supply system. The same does not necessarily apply to car drivers. Here exposure depends on the exhaust of the vehicle in front, and the air pollution that drivers breathe in is often worse than that of a pedestrian. We are fooling ourselves if we think that we can hide in our cars—and by hiding in our cars we are making the pollution problem worse.

Many cities already provide air quality information to their people in the form of an index. This tells residents how good or bad the air is. It is this type of data that catapulted Beijing's air pollution onto the front pages of newspapers around the world. Most indexes also give health advice, which generally consists of telling vulnerable people such as the elderly or children to avoid outdoor exercise when pollution is bad. Air pollution varies by time of day, so changing the time that you go for a run or a walk or moving summertime school sports to the mid-morning rather than afternoon may help to reduce exposure. There is little evidence, however, that large numbers of people heed the current warnings and change their behavior.[7] This is understandable. Is it fair to only ask those who suffer most from air pollution to change their lives? Why should people whose lives are already being affected by air pollution be asked to compromise further still? It would seem morally unjustifiable. The responsibility should surely lie with polluters to change their behavior and not the victims.

If we could each measure the air pollution around us maybe we would be conscious of the problem, but reliable air pollution measurement has always been a highly technical and scientific process. Huge excitement has therefore been created by the promise of small sensors allowing each of us to measure the invisible air pollution that we breathe. This would mean that we could avoid polluted places, change our polluting behaviors and put pressure on our leaders for solutions. Such sensors can now be bought online for less than a hundred dollars. Simply visit an internet site, pay your money and the sensor will arrive in the mail. The results from people's houses and gardens are starting to be posted on the internet.

These sensors have not been enabled by new scientific discoveries but by coupling better computing power with technologies created in the late twentieth century. Sadly, they do not work as well as we would hope. Many sensors use technologies that are effective as alarm systems in factories and laboratories, but they struggle when faced with the challenge of measuring air pollution that we breathe outside. Sensors designed to measure a single pollutant often perform badly when they encounter the cocktail of different pollutants that we have in our outdoor air and the trace quantities involved. They can also be confused by the rapidly changing humidity and temperatures that are part of our everyday weather and as we move from indoors to outdoors. Inaccuracies can be huge. Some of the leading independent tests on sensors have been made by Ally Lewis,

Pete Edwards and their team at the University of York.[8] They bought some sensors and put them through their paces in their laboratories and on the roof of their building at the edge of the campus. They took identical sensors straight from the box, as you or I would buy them. Some sensors measured more than six times as much ozone as others. Some sensors would double (or halve) concentrations as humidity changed and the readings from some carbon monoxide sensors drifted by around 30 percent in a month. They also found that the carbon dioxide from traffic could swamp measurements of nitrogen dioxide. This is worrying and runs the risk of misleading users, falsely alarming or falsely reassuring them. It is unclear if greater uptake of the current generation of small sensors will inform people about air pollution or flood them with questionable data.

Images of the polluted streets of East Asia often show people wearing masks. Some trials on the streets of Beijing have shown that if you have a heart problem, a mask can help reduce symptoms, but key to the effectiveness of a mask is the fit around your face. Even a small leak from having stubble, facial wrinkles or a beard could render the mask useless. Less explored are the negative effects from having to breathe harder when wearing a mask, which could place additional strain on the heart and lungs of people already in bad health. There is little evidence on how to balance these negative effects against any positive impact, and certainly telling people to avoid doing things outdoors needs to be balanced against the positive effects that exercise brings.[9]

So, if we can neither clean the polluted air nor easily avoid exposure, we have to stop the air being polluted in the first place. Industry often objects to new regulations and standards, so one solution is to impose minimum environmental standards internationally. Examples include the pollution control directives that apply across the whole European Union. Although frequently labeled as red tape, these common regulations mean that industries cannot undercut one another by moving from country to country and thereby transferring the cost of their pollution impacts to the state or wider society.

The challenges of reducing air pollution from industry are small compared to the difficulties of tackling air pollution from over 1.3 billion vehicles used around the world,* but at least these are mainly registered in some way, must meet minimum standards when they are made and are renewed every ten years or so. As we saw in Chapter 11, a far greater problem still is controlling the air pollution from cooking and heating in the many billions of homes worldwide.

One quick solution is to improve the quality of the fuels used. Several examples have already been discussed earlier in this book. They include smokeless fuels and natural gas to solve the coal smoke problems in UK cities; the removal of sulfur from road fuels in Europe, which led to dramatic reductions in the number of particles in urban air;[10] and of course, the removal of lead additives from gasoline. Perhaps the greatest success story relating to smokeless fuels comes not from London but from Dublin. In 1982, Luke Clancy, working as a doctor at St. James' Hospital in Dublin, was dealing with a serious crisis.[11] An extra fifty-four inpatients had died that January compared with the years before. Clancy had to find out why, and quickly. Despite extensive investigations, no bacterial or viral cause

* Data for 2015. For country by country statistics see http://www.oica.net/category/vehicles-in-use/.

could be found. Clancy was befuddled until he looked out of the window and saw the smoke rising from the chimneys of homes across the city. Maybe the problem was not inside the hospital but outside.

Coal-burning in Dublin had increased a great deal during the 1970s in response to increasing oil prices and inducement from government grants. Clancy contacted the city council to obtain air pollution data. As in the 1952 London smog, the smoke and sulfur dioxide peaks matched the increased deaths. It was hard to know how many people were dying as a direct result of the coal-burning but it was clear there was a serious problem and that quick action was needed. Rather than following the UK route of declaring smoke control areas and upgrading home fires and boilers, the city council simply banned the sale, marketing and distribution of smoky (bituminous) coal. This required people to burn smokeless coal or other fuels instead. It was an instant success. The effect was dramatic.[12] Wintertime black smoke decreased by 70 percent compared to the pre-ban period; deaths from respiratory problems dropped by 16 percent and cardiovascular deaths by 10 percent, nearly 360 fewer deaths per year. Many people gave up on solid fuel use altogether and swapped to natural gas heating as this became available in the city. Following the Dublin experiment, the smoky coal ban was rolled out to another eleven Irish cities, resulting in reductions in black smoke of between 45 and 76 percent.

As with the Six Cities study, the initial data was completely reanalyzed by a later team of scientists.[13] Rather than looking at health data before and after the bans, they compared towns and cities where the new coal bans were in force with areas that had no bans. With another seven years of data and more cities for comparison, the reanalysis gave a different result to Clancy's original work. Cardiac deaths had dropped everywhere, not just in the areas with the bans. This was most likely due to the huge economic, social and health care improvements in Ireland during the economic boom that started in the 1990s. However, it was also clear that the reduction in deaths caused by breathing problems in Dublin, Cork and four other areas were linked to the bans. This is one of the few global cases where a policy to improve air pollution could be tracked by measuring a near-instantaneous change in city air and the positive impact this had on health.

Lastly, and most often overlooked among our pollution sources, is agriculture and, more broadly, the way in which we manage land. Many of us think of the countryside as a low-pollution environment and escape into its rolling green hills for a respite from the city, but farming is an important source of particle air pollution (see Chapter 6). The burst of ammonia released from fertilizing crops, spreading manure stored overwinter and moving animals to outdoor summer grazing all play an important role in the particle pollution that plagues Europe each springtime. This is becoming an increasingly important problem across the developing world too—especially in East Asia, where pollution from new, badly controlled industry mixes with ammonia from agriculture and increased meat production. In the United States, the control of ammonia from farming would now be the most cost-effective way to reduce regional particle pollution. A decrease of 50 percent in ammonia pollution from farming could reduce the annual global mortality from particle pollution by around 250,000 people, including around 16,000 in North America, 52,000 in Europe and 105,000 in East Asia.[14]

We cannot stop farming. We need food to eat, but simple measures such as covering slurry stores, installing better housing for animals and injecting fertilizer into soils rather than spraying it in the open air would go a long way toward achieving these targets. The role of agriculture in our air pollution is clear, but it is badly understood by farmers. It is also badly understood by governments who have spent decades developing control measures for transportation and industrial air pollution. This means that only modest reductions of 6 percent in EU ammonia are planned from 2020 to 2030. This contrasts with much more ambitious decreases of almost 60 percent in sulfur emissions and nearly 50 percent in particle pollution.[15]

Ammonia is not the only air pollutant derived from farming. In many parts of the world, fields are set on fire to clear stubble, weeds and waste before sowing new crops. While burning may be a fast way for farmers to clear their fields, it is highly unsustainable. It produces large amounts of particle pollution and can lead to reduced soil fertility, making farmers more reliant on costly artificial fertilizers. Agricultural burning was a major factor in the 2016 and 2017 smogs that engulfed Delhi and the surrounding region. Traditionally the annual burning of the remnants of rice crops took place in late September and October. In 2009, a new law delaying the planting of rice to improve soil conditions shifted waste-burning to a time when seasonal winds carried the pollution to the Delhi region.[16]

We often see dramatic coverage of wildfires and forest fires in news reports. These are the fires that come close to where people live in the developed world, mainly across North America and Australia. Sydney's population of four million people have grown used to the smoke from planned fires in the Blue Mountains to the west of the city. These led to an increase in people being treated in hospitals for breathing difficulties, especially among those with problems such as asthma.[17] Europe has problems with landscape fires too. Fires in Spain and Portugal added to air pollution problems across Europe during the 2003 heat wave and contributed to the "red sun" events of 2017, when smoke in the upper atmosphere masked the sun with a red haze and a dense brown cloud rendered central and southern England in sepia.[18] Streetlamps switched themselves on, cars needed headlights and lights were required indoors during the day. Social media and news reporters likened the effect to scenes from *Blade Runner*, but it showed how far air pollution from wildfires can travel.

This is not the only example. Agricultural and forest fires in Russia in 2002 and 2006 also caused air pollution problems across Europe as far west as the UK.[19] Global air pollution from such fires has been estimated to cause around 330,000 early deaths in an average year.[20] Around half of these fires happen in sub-Saharan Africa, with a further third in Southeast Asia, where Indonesian peatland fires are the biggest source. We hardly ever see these reported in the news. It is wrong to think of such fires as being natural. Although some grass fires in the African savannah are part of a natural cycle, other global wildfires are usually not natural at all. They often occur in managed landscapes and are linked to agriculture or land clearance for crops. Peatland fires in Malaysia and forest fires in South America rarely start without human intervention. In a worst-case El Niño year, Indonesian fires can push the global total of early deaths to over half a million people. Particle pollution from land clearance and fires in South America adds around 10,000 premature deaths in an average year.

So, successes in battling air pollution have mainly come from improved fuels or from technology to control pollution where it is produced, rather than from eye-catching attempts to clean our polluted air. Necessity is the mother of invention and each week I get emails from innovators with new ideas to make money from scrubbing our city air. These range from networks of pipes that suck air from polluted places to wonderful ways of installing greenery on benches and walls, complete with extraordinary figures for the tons of pollution each year that even small installations of urban vegetation can absorb, in defiance of the laws of physics. The creativity and conspicuous design in some of these solutions is to be celebrated and is far greater than, say, a dull metal box of cleaning technology hidden inside the engine compartment of a bus. Schemes that attempt to clean the air risk diverting resources from controlling pollution at its source. However, they can be politically attractive. For example, green walls provide very conspicuous evidence that money is being spent on pollution control but, to paraphrase Simon Birkett of the Campaign for Clean Air in London, focusing on small solutions with little impact risks us becoming "busy fools" to scant effect.

We also have to remember that air pollution is not caused only by transportation or solid fuel burning but is created by many other sources, including agriculture, that need to be addressed too. These may be some of the toughest battles, for the simple reason that, like wood-burning, everyone thinks that farming is merely managing a natural resource. There is little knowledge that it can produce air pollution every bit as harmful as the traffic on our streets.

Part 4

Fighting back: The future of our air

Chapter 14

A conclusion: What happens next?

We stand poised on the cusp of great change. The world's infrastructure is predicted to double in around fifteen years, and the numbers of people living in towns and cities will double in the next forty.[1] An extra two billion urban residents will need places to live, new services and ways to move around their cities.[2] Over 80 percent of global GDP is generated in cities, and urbanization can contribute to sustainable growth if managed well. At the same time, we have to reduce the health burden of air pollution. No country is pollution free and very few cities meet World Health Organization standards. The decisions that we make now, at the start of the twenty-first century, will determine the everyday lives of the next generations.

There would be an outcry if London's tap water was killing 9,400 people a year. International businesses would relocate, and the tourists would stop coming. The UK's reputation would be in tatters and London would dwindle as a leading world city. Heads would roll in the boardrooms of the utilities companies and ministers would lose their careers if they allowed the situation to continue year after year. So why should we treat the consequences of air pollution any differently? The maximum estimate of 9,400 early deaths[3] each year from air pollution in London has been correctly labeled a public health crisis. The scientific evidence that air pollution is harmful to health is overwhelming, but scientists do not make policy. It is the job of political leaders to determine the action that is taken and balance objectives and needs.

Despite the magnitude of this challenge, air pollution is largely unrecognized in our political debates. Reducing the air pollution burden and creating a healthy environment for our children is not a regular feature of elections and gets little mention in the visions of the future painted by our politicians.

A notable exception from the past was Willy Brandt, who in April 1961 was campaigning to become Germany's Chancellor. He was already mayor of West Berlin and the election campaign was a bruising one, with the Christian Democrats using Brandt's illegitimate birth and wartime emigration against him.* Speaking on the campaign trail at the Beethoven Hall in Bonn, he boldly laid out his new environmental vision demanding that "the sky over the Ruhr [area] has to become blue again."

Cleaning the air of the heavily industrialized Ruhr was considered impossible. Over 300,000 tons of soot and dust fell to the ground each year over the region. Air particle pollution during smogs was worse than anything seen in Beijing in the early

* Brandt was defeated on this occasion but became Chancellor in 1969. His obituary can be read at http://www.independent.co.uk/news/people/obituary-willy-brandt-1556598.html.

twenty-first century[4] and the polluted air was having a clear impact on the people and the environment. Death rates increased after each smog and children suffered from bronchitis, rickets and conjunctivitis. They also tended to weigh less than children from other parts of Germany. The cattle raised in the region weighed less too.

The cause was the eighty-two blast furnaces, fifty-six steel convertors and ninety-three power plants that operated almost entirely without pollution controls. Workers had to routinely remove thick dust from factory roofs, and as if this was not bad enough, home coal-burning added to the pollution burden. Writing in 2015, the academic Anna Lisa Ahlers described her father's memories of growing up in the Ruhr:

> I went home after playing outside, I was always covered in dust and soot. My classmates and I suffered from chronic bronchitis throughout our entire childhood. And trying to dry [the washing] was a real headache: when you put it outside, you know, it was always coated with a grey or even black layer.[5]

Air pollution was seen as an inevitable consequence of industry and seemed impossible to change. Willy Brandt was therefore ridiculed for his "blue skies" idealism. However, his speech resonated. Brandt catapulted a previously ignored regional issue into mainstream debate and drew attention to the health price that people were paying for Germany's economic miracle. In the years that followed, state regulators took on vested industrial interests, and won. New laws, and importantly new attitudes, came to the fore and

Brandt is now credited with sowing one of the seeds of the German environmental movement.[6]

The use of the air as a means of waste disposal meant that the Ruhr's massive industrial growth imposed a heavy price on people's health. This has been a common thread throughout the history of urban air pollution, starting with the burning of coal in medieval London, through the industrial revolution and the evolution of our cities until the more recent growth in motorized transportation. The rise of China's industrial might and its subsequent air pollution problems is another rerun of this familiar story. Managing the health and environmental impacts of air pollution requires us to break this cycle.

With the Trump administration focusing on "America first" and Europe questioning its identity, it seems inevitable that world development in the next decades will be shaped by China, and it will probably come from Xi Jinping's Belt and Road Strategy. This is set to be the most ambitious infrastructure and development program that humanity has ever seen. Covering seventy countries and around 65 percent of the world's population, it will create land and sea routes, along with energy infrastructure, westward from China across Asia to Southern Europe and East Africa, moving a quarter of the world's goods.[7] It is an unparalleled opportunity for investment in green and low-pollution development, but as ever this depends on political priorities. China is changing at home. The smogs that beset Beijing became intolerable in terms of their damage to the health of the people and China's international image. Action was taken on home heating and industry. Air pollution in the sixty-two Chinese cities tracked by WHO dropped by an average of 30 percent between 2013 and 2016. If these changes are continued and spread westward by the Belt and Road strategy, the life quality and air pollution experienced by billions of people could be transformed. This will not happen without political leadership and depends crucially on Xi Jinping's vision for China and South Asia.

Future clean air needs leadership and vision on many levels, not just from national governments but from those who lead our cities. Sadiq Khan became mayor of London in 2016 and prioritized the need to reduce air pollution, using it as the topic of his first policy speech and setting a new direction for the city government. Speaking at Great Ormond Street Children's Hospital, he compared the modern challenges to those that had led to the Clean Air Acts sixty years before:

> British politicians at the time did an amazing thing and responded on the scale that was required. Today we face another pollution public health emergency in London and now it's our turn to act for the good of Londoners and for future generations to come . . .* Just as in the 1950s, air pollution in London today is literally killing Londoners. But unlike the smoky pollution of the past, today's pollution is a hidden killer.[8]

Two years later, electric taxis and buses are plying London's streets, the world's first low emission zone for construction has been created, the city's low emission zone for traffic is being tightened

* I think the term "public health emergency" was first used by Simon Birkett of the Campaign for Clean Air in London.

and plans are being created to ensure that major new developments contribute to reducing London's air pollution.

In his Great Ormond Street speech, Khan emphasized the importance of managing air pollution inside cities. Mayors and city leaders need to consider the health of their residents, but the importance of managing urban air pollution goes far beyond this. Air pollution from factories does not stop at the factory fence. The Chinese megacities of Beijing and Tianjin, as well as Karachi in Pakistan, cause more harm to the surrounding population than to their own residents. For other, more typical megacities, the impact of particle pollution on people downwind can still be around 40 percent of that experienced by city residents. If the pollution resulting from chemical reactions downwind is considered, then the impact on surrounding areas is greater still.

Improving air quality makes economic sense too. In 2011, the US Environmental Protection Agency looked back at its Clean Air Act program for the thirty years between 1990 and 2020. An investment of $65 billion produced a whopping benefit of $2 trillion. The benefits from pollution reduction were thirty-two times greater than the costs. Cleaning the air was a "good investment for America."[9] In Europe, cleanup investments in the energy, industrial and road transportation sectors were saving an estimated 80,000 premature deaths in 2010. An early death, especially of a young person, is a tragedy but also a loss to the economy. Better air pollution was estimated to have saved Europe the equivalent of 1.4 percent of GDP per year.[10]

Solving our air pollution problems seems a massive task. It feels too big for any of us to make a difference, but we all need to be part of the fightback. Air pollution is caused by the way that we live our lives, yet simple changes can make a big difference.

The greatest opportunities are in the way that we travel. Reducing the traffic in our cities will do so much more than reduce air pollution. As discussed in Chapter 12, 40 percent of car journeys in England are of less than two miles and 60 percent are less than five miles.[11] The gains from leaving the car at home and walking, cycling or taking public transportation for these short trips would transform our cities. It would tackle air pollution, traffic noise and climate change emissions, and combat the increasing burden of disease resulting from our inactive lifestyles. We can all do this in some way. As you step out of your house and into your car, ask yourself if you could walk instead. The times spent walking my daughter to school, skipping to the next lamppost and counting our steps, are memories that I will always cherish. This set her up to walk herself to school, friends, clubs and shops as she became older.

For many short city journeys, walking and cycling can take less time than it takes to drive and park our cars. If your town or city was built more than a hundred years ago then it was made for walking. Using the local shop rather than driving to the out-of-town supermarket would rekindle our town centers and reduce community isolation. Many will say that this cannot be done, but the evidence shows otherwise. Countless Danish and Dutch cities have rowed back from domination by cars to make their centers people-friendly places and, in 2016, the bicycle became the dominant rush-hour vehicle in the City of London.[12] If traditionally conservative bankers and financiers can be persuaded to swap their bowler hats for cycle helmets, then this can be done in your town too. We can turn our streets back into places for living, rather than places for driving. We need to make being car-less or car-lite the new normal.

There is abundant evidence that building more roads does not ease congestion or make travel any easier. New roads become clogged with new traffic. Fortunately, the reverse is also true: unmaking roads can reduce the traffic in our cities. Imagine how your city would be transformed if the roads in your neighborhood became parks for people rather than parking for cars. It can be done. A great example comes from Seoul in South Korea. Between 1973 and 2003, a three-and-a-half-mile, four-lane elevated expressway took traffic right into the heart of the city.[13] One hundred and seventy thousand cars used the road each day and it was frequently congested. Instead of building more lanes, the city authorities demolished the whole expressway. Doubters predicted traffic chaos but traffic in the city center decreased. The residents of Seoul adapted the way that they traveled, many swapping to the subway. Part of the vision for the city was to restore the lost Cheonggyecheon River, which had been buried under the road. The former expressway is now a long riverside park with 1.5 million trees. Insects, birds and fish have returned and it is a hugely popular place for the people of Seoul to relax. It is a tourist attraction, with thriving businesses, a venue for festivals as well as a cycle route. This vision for low-pollution cities has to apply in both the developed and developing world.

Sadly, this vision is not at the forefront of government actions to combat air pollution. So far, we have not paid sufficient attention to reducing the underlying causes. The ongoing battle to control air pollution from traffic has seen huge innovation but we have expended less effort in reducing the growth in motorized transportation, eschewing the views of those who are keen to embrace a new vision for their cities in favor of those who only view transportation through a car windscreen.* In the same way that new roads lead to more traffic, new cycle infrastructure, easier ways to walk and improved public transportation encourage new ways to travel.

Similarly, increasing the efficiency of home heating and insulating houses was not part of the solution to London's smogs. It could have played a vital role in reducing air pollution and would also have reduced the high winter death rates that continue to affect the UK. Rather than find ways to use less electricity, we moved power stations to the countryside, built tall chimneys and created the acid rain problems that came to the fore in the 1970s. The list could go on.

A new battleground has come to the fore: our homes. Abundant evidence shows that home heating with solid fuels, be it wood, coal or peat, is disastrous for urban air pollution. Even in small towns, a few homes using solid fuel can be the dominant source of particle pollution. This requires us to question our own attitudes. Is it right that the cozy fire in our living room is harming our neighbors? We need to seriously question if home wood-burning should be permitted in the towns and cities of the twenty-first century when we have much cleaner alternatives to heat our homes: gas central heating, heat pumps, electricity or district heating. Across parts of the developing world there are fewer fuel options and the home use of solid fuels leads to millions of early deaths each year, with the young very badly affected. Ending this tragedy is an important international

* Thanks to Philip Insall, formerly from Sustrans, who explained how the so-called windscreen view dominates transportation thinking.

development goal, but it will not be easy. Better cooking stoves have helped not only to improve air quality but also to free girls and women from the daily chore of collecting firewood, giving them time for education or earning for their families, but the solutions need to focus on economic and infrastructure development rather than just replacing fires with stoves.

The new home battlefront includes more than heating and cooking. In 2018, it was shown that the pesticides, coatings, printing inks, adhesives, cleaning agents and personal care products used in the United States now dominate the pollutants that form ozone.[14] The finding will be equally applicable across Europe and beyond. Manufacturers will say that their responsibility ends once the product leaves the factory, but that will do nothing to solve the problem. Control of air pollution may end up with the choices that each of us make in the supermarket and on main street and when we buy online. This will require better product labeling and information for us as consumers, or restrictions on the solvents that can be used.

Despite the overwhelming economic benefits of investing to reduce air pollution, policies and strategies are constrained by what is thought to be politically palatable. This means that many strategies needed to resolve our air pollution problems never find their way into the discussion, let alone into city plans or government policies. This is especially the case for transportation, where industry lobbying is strong. The voice of industry is always louder than the voices of the people who suffer from poor air. Shifting this consensus is a job for all of us. We can do this by our own actions and normalizing low-pollution lifestyles. There are little things that we can all do. Share the school trip with other parents and walk the kids together; plan your jog to take the back streets and reduce your exposure; if you have to drive then combine your car journeys together; shop local. These are just a few examples. We can all join the debate too and help to shape our future. Neighborhood groups, residents' associations, environmental groups, cyclist groups, school parents' associations, the letter pages of newspapers and political parties are all places to bring about change. Trade unions are growing increasingly concerned about the air pollution that outdoor workers breathe. Contact your political representatives too; letters to government are all read, and sometimes replied to. Campaign groups have achieved great successes in the past. The *Los Angeles Times* led a campaign that pressured city government to tackle the eye-stinging smogs of the 1950s, and the environmental movement in Germany ensured that images of forest dieback from acid rain appeared in newspapers around Europe. In the UK, a public campaign caused an unprecedentedly swift government response to the Royal Commission's finding on lead in gasoline. At city level, Simon Birkett of the Campaign for Clean Air in London worked tirelessly to keep air pollution on the London mayor's agenda and paved the way for tightening London's low emission zone. The environmental lawyers ClientEarth have opened up a new battlefront using the law to defeat the UK government on the adequacy of its air pollution plans, not once but three times since 2011. Each time the government was forced to improve their air pollution plans, to bring about improvements faster and to take more action in more cities. It remains to be seen what will happen to these plans once the threat of EU sanctions is removed from the UK government, post Brexit.

The "polluter pays" principle should be enshrined in law, making industry responsible for the pollution that they cause and for the harm caused by their products. Businesses also need to take responsibility for the air pollution created by heating and cooling their buildings, their employee travel and their deliveries. Many businesses are starting to hear this message. I am constantly encouraged by the retailers and office owners who contact me concerned about the impact of air pollution on their workers and customers, and on their reputations. There are countless examples of businesses that are trying to reduce their air pollution footprint by changing the way that they distribute their goods and the way that their employees travel.

Ports and airports are being recognized as important sources of air pollution, but it is hard to regulate the boats and planes that travel between nations. Ships that sail close to many of Europe's shores have to burn reduced sulfur fuels, but air pollution from international travel is often viewed as another country's problem, meaning that no one takes responsibility. Taking the train rather than flying is a simple way to reduce your air pollution footprint. Several chapters of this book were written on train journeys around the UK, and to Spain and Switzerland. It provides a great perspective on a journey and is often faster for trips between city centers, but the tax breaks given to the airline industry mean that millions of people will continue to be blighted by their noise and air pollution.

Science has a vital role to play in the battle for clean air. This book has covered many of the mistakes of the past, where actions were taken without looking at the whole evidence base. Scientists need to explain their evidence so that the problem is understood and the solutions are guided by what works. It is not sufficient to rely on ministers, civil servants or policymakers to read our publications in scientific journals or see our presentations at air pollution conferences. We need to be speaking at public meetings, engaging with the press and talking to politicians.

There is, however, a fine line between explaining the evidence and being dismissed as a campaigner with a partisan view. I am lucky to have worked alongside scientists who understand that communicating their work is an important part of their role. To communicate effectively we need to establish a terminology that allows complex evidence to be put in clear and simple terms. The first health burden calculations for the UK, which expressed the impacts of particle pollution in terms of 29,500 premature deaths in 2010, was a major advance in explaining the consequences of our air quality problems. But one major issue is that scientists make their living from doing research. We entered science to provide answers to questions. Too often scientists focus on what is not yet understood, rather than on the overwhelming evidence that air pollution harms our health.

It is not just air pollution scientists who need to speak out. The tobacco debates took an important step forward when doctors told people to stop smoking in 1962.* The 2016 air pollution report by the Royal College of Physicians was therefore an important landmark, with doctors highlighting the harm that air pollution does to their patients and calling on government to take action to properly address the issue.

* Before the report, between 1956 and 1960 UK government spent £5,000 on educating the public about the risks from smoking; £38 million was spent on cigarette advertising. See Royal College of Physicians, *Smoking and Health*. London: RCP, 1962.

Industry needs to get onside too. The use of lead additives in gasoline taught us the important difference between no evidence of harm and evidence of no harm. The battle for clean air has been obstructed at every turn by those who profit from using our shared air to dispose of their waste. There are, sadly, many examples in this book where industry has fought tooth and nail in the face of inconvenient truths. Recently we have seen car makers develop diesel vehicles that pass pollution tests but behave completely differently when used on the roads. The manufacturers have yet to justify this to the European public. Governments need to ensure that regulations prevent industry from cutting corners, ensuring that it makes good economic sense for business to cut air pollution. I look forward to products being sold on the basis of their low air pollution emissions.

When actions have been taken, governments often focus on single sources or single pollutants rather than all sources of air pollution. We thought the UK's air pollution was solved when London's winter smogs were banished, but air pollution from road transportation and acid rain crept under the radar. Similarly, wood-burning returned to the cities of northwest Europe while we battled with the air pollution from traffic. We need a more holistic view of the way in which we use our air as a waste disposal route. Politicians need to think of air pollution as an opportunity to improve public health rather than as an endless series of problems that are hard to resolve.

The mobility of our atmosphere means that our waste is taken away to where it is no longer our problem, but one of our mistakes has been to confuse dilution with rendering our air harmless. They are not the same. Another mistake has been to ignore cumulative impacts. The impact of one fireplace is small, but millions of coal fires together killed thousands of people in the London smog of 1952. The combined emissions from power stations, factories and homes led to the acid rain that damaged the forests of northern Europe and North America. A few grams of gasoline additive in the fuel tanks of billions of cars resulted in lead becoming a ubiquitous health-damaging global pollutant.

We have also failed to see how effects accumulate over time. The smogs of the 1950s and 1960s showed how short periods of severely polluted air could harm the health of thousands of people, but it took another forty years before the Six Cities study revealed that breathing polluted air every day was shortening our lives. This finding has been followed by evidence that children's lung growth can be stunted by the traffic pollution that they breathe daily. New research is now pointing to lifelong impacts of air pollution from pre-birth, through our development as children and on to our eventual lifespan as adults. The history of air pollution is littered with warnings that went unheeded: Donora and the Meuse Valley, lead in gasoline and the problems with diesel cars. We would be reckless not to listen to the modern-day warnings; we ignore them at our peril.

Bans on indoor smoking transformed a night out in a restaurant or pub and made us realize just how bad it had been to eat and drink in a smoke-filled space. It seems inconceivable that this would ever be reversed. Indoor smoking bans in ten countries, twelve US states and fifteen cities around the world led to an average 12 percent decrease in heart attacks, along with reductions in strokes and in children's asthma. No one expected this. It showed that passive smoking had effects that we never realized before.[15] Reducing our outdoor air pollution might lead to even greater health gains than we can imagine.

There are huge social injustices at the heart of our air pollution problems. In 2011, I gave a public talk in a church hall in my home city. I showed a map of car ownership and a map of air pollution. The cars were mainly owned by those in the wealthy suburbs, but they experienced the least air pollution. The worst air pollution was in the city center and alongside the busiest roads, where car ownership was lowest. The people who lived with the worst air pollution were not the cause of the problem; they mainly walked, cycled or took public transportation to work and school. This pattern is repeated throughout the UK and the developed world. Internationally it is the world's poorest who suffer the greatest air pollution burden: those living in a crescent across Africa and through to Southeast Asia. The people with least access to food are suffering the worst damage to their crops. Relieving the burden of air pollution needs to be a target for international aid and placed firmly and explicitly within sustainable development goals.

Our air is not owned and cannot be owned. Its mobility means that the air I breathe in southern England today might have been in Paris yesterday, and tomorrow it could be in Amsterdam. Our air is the ultimate shared resource, but this has not encouraged us to take additional responsibility for its care; it seems that the opposite is true. In his 1968 paper "The Tragedy of the Commons,"[16] the biologist and economist Garrett Hardin[17] drew upon a pamphlet and lectures first delivered by the Victorian economist William Forster Lloyd, who discussed the fate of an English Common—a piece of shared land where all members of the local community could graze their animals. Lloyd and Hardin considered what happens if someone puts an extra cow onto the common. It eats some of the shared grass. The burden of the extra cow on the common is shared by everyone in the village, as all the animals suffer from the marginal effect of the overgrazing, but the profit goes to one person—the owner of the extra cow, when it is sold at the market. It is therefore in the best interests of each villager to put extra cows on the common to make more profit for themselves, spreading the cost burden between them all. Similarly, there is a cost for a polluter to clean their vehicle exhaust or factory emissions, but their share of the health and environmental burden is tiny. It is only when we look at the whole system that we see that the total cost of cleaning our air is far smaller than the consequences of not cleaning it.

The smoking chimneys of an industrial town used to be regarded as a sign of prosperity. Even today, pollution means profits. The challenge for politicians is to make the economics work, to ensure that the polluters pay and that it is rational to act in the right way. Hardin discussed the morality behind abuses of our common resources—the social pressure put upon the farmer who grazes more than their fair share of cattle—but it is difficult for a parent of an asthmatic child to apply social pressure to the thousands of motorists who drive past their home. Our air pollution problems can only be resolved through collective actions across society, which brings us back to actions by government.

The title of this book frames air pollution as an invisible killer. No one has air pollution listed on their death certificate, but there is overwhelming evidence that air pollution is shortening our lives. It increases deaths and illness from everyday causes, including respiratory problems, heart disease, strokes and many more. We can see air pollution if we look for it. We can taste and smell the smoke from a fire and the exhaust from a passing car. The thick winter smogs that affected London and the haze in cities around the world today are visible signs of air pollution, but we no longer perceive them; they are part of everyday life. During the 1921 coal strike, people in the UK saw that the world around them was miraculously transformed; distant hills were visible in a way that people had never seen before. During the Asia Pacific Economic Conference (APEC) in 2014, and the 2015 parade to mark seventy years since the ending of the Second World War, industry around Beijing was curtailed and traffic cut in half. The air pollution decreased, and the skies cleared. The people of Beijing could see the true color of the sky without looking through the customary haze. It was nicknamed APEC blue, and later parade blue.[18] It is not the invisibility of air pollution that is the problem but its normalization and acceptance.

The acceptance of air pollution extends beyond our everyday perception to the views of governments. In 2005, EU leaders decided on a policy that would still include a projected 200,000 early deaths per year by 2020.[19] Their later plans for 2030 similarly set aside the recommendations from health experts and economists.[20] They accepted air pollution as the norm and that bad would be better than very bad. Instead it is clean air that should be our benchmark.

This book has allowed us to look backward in order to look forward. The history of air pollution is littered with early warnings that were never heeded and abrupt changes in government policy in response to disasters that could, and should, have been avoided. It took the deaths of 12,000 Londoners during the winter of 1952 for government to heed warnings from the decades before. Lead was a global pollutant affecting millions of children before it was controlled, and China's annual air pollution death toll reached 1.7 million before the problem became a national priority.[21]

Many targets have been set for the decades to come. The UK aims to meet the 2010 legal limits for nitrogen dioxide by the mid-2020s. The probability of achieving these depends crucially on decreasing exhaust emissions from diesel vehicles that we will buy in the next few years and how they perform when driven on our streets. Based on the past track record it is hard to be optimistic that these targets will be met, and at current rates of progress, nitrogen dioxide alongside our city center streets and main roads will be still be exceeding legal limits for many years or even decades to come. This is why new policies and actions, including the proposed low emission zones in our major cities, are so important.

European limits for particle pollution have been criticized for not being sufficiently protective. For nitrogen dioxide the legal limits match the World Health Organization guidelines, but for particle pollution the limits are much less strict and were set to be twice as high as WHO recommended. Debate is therefore turning toward targeting WHO guidelines for future policy. In 2018, the UK government consulted on plans to reduce particle pollution and halve the population living in areas with airborne particulates above WHO guidelines by 2025. These areas are mainly in the highly populated southeast, including London where the mayor wants to meet the guidelines by 2030. This will require us to move beyond the recent focus on traffic and industry to manage our air more holistically, including addressing areas such as home wood-burning, agriculture and catering.

The 2015 Paris accord set out a program to limit global temperature rise to 2° C by the end of the twenty-first century.[22] The next decades will therefore be defined by battles over climate change. We will be fighting to reduce our climate change emissions and to maintain our lives and economies in the face of rising sea levels and changing weather patterns. Coal has fueled our industrial and social development for nearly three hundred years, from the industrial revolution in Europe to the recent industrialization of China. But this has come at a price. Air pollution from the burning of coal features throughout this book and no other human act has had such a profound impact on our planet. Coal-burning has damaged our health and curtailed the lives of millions of people, and it continues to do so. It has damaged ecosystems and has been the main contributor to the increased carbon dioxide in our atmosphere. Climate change science has finally allowed us to reframe coal from being a valuable natural resource to a dangerous substance that must be left in the ground if we are to prevent uncontrollable global warming.[23] The curtailment of coal-burning can only be a good thing for our air, especially if it is replaced by energy harvested from renewable sources such as hydroelectric, geothermal, wind, solar, wave, tides and, as some advocate, nuclear power.

In the twenty-first century it will be even more important for us to consider air pollution alongside climate change, so that actions to reduce one do not worsen the other. As recognized by the World Health Organization, there are great benefits from balancing the right policies that tackle both air pollution and climate change at the same time.[24] Many air pollutants, such as black carbon and methane leakages from natural gas and coal mines, are bad for air pollution and for climate, and also lead to food crop harm.[25]

In 2008, the UK government adopted the Climate Change Act. This law required a reduction of 80 percent in carbon dioxide–equivalent emissions by 2050, compared with what the UK emitted in 1990. Between 2012 and 2015, emissions fell by an average of 4.5 percent a year but this was almost entirely due to progress in electricity generation, reduced coal usage and an expansion of electricity from natural gas, wind and wood-burning.* There had, however, been almost no progress in the rest of the economy when it came to reducing greenhouse gas emissions.

Clearly, there are many ways to bring about the required changes. In 2018, Martin Williams and colleagues at King's College London looked at the UK's future energy plans and their air pollution implications.[26] There was some good news. A reduction in fossil fuel use will lead to decreased formation of particle pollution over the whole country and, with our European neighbors following the same pathway, springtime in Western Europe in the 2030s will no longer be punctuated by severe periods of particle pollution. But for those of us who live in cities (80 percent of the UK population) there was some bad news in Williams' projections too. The two scenarios that meet the requirements of the Climate Change Act would have serious implications for urban air pollution, mainly from increased wood-burning for heat and energy in the next two decades but also from the widespread use of combined heat and power, which increases urban nitrogen dioxide. Instead of reducing

* Burning wood chips for power generation is recorded as zero carbon in the accounting systems for greenhouse gas emissions. Reality is not so simple. See Chapter 11.

road transportation, following the trends of the early twenty-first century, future policies include planned increases. These will make the challenges posed by carbon and air pollution more difficult to deal with.

The technologies of the future are often waiting in the wings, yet to be widely adopted. Many national and city governments are committing themselves to banning the sale of conventional gasoline and diesel cars, in the case of the UK by 2040. It seems inevitable that gasoline and diesel cars will be replaced by emerging battery electric vehicles. The cost of battery packs fell by 80 percent between 2010 and 2017,[27] but electric cars require investment in charging infrastructure and changes to our electricity generation. We must also remember that electric vehicles will not mean air pollution is reduced to zero. They still produce particle pollution from the wear on brakes, tires and roads.

A greater challenge comes from powering long-distance heavy goods vehicles, where currently diesel is the only viable option. Alternative fuels such as methane pose large climate risks and hydrogen-fueled trucks are not yet in production. It therefore seems inevitable that diesel will be with us for decades to come, and far greater investment is needed to ensure that it is as clean as possible. Similar challenges are posed by the lack of alternatives to fuel oil for shipping and kerosene for aircraft, making electric trains a viable low air pollution and low carbon alternative for the long-distance transportation of people and goods. Autonomous or driverless vehicles are also under test. While these might increase mobility for some, they risk drawing people away from active travel and high-occupancy public transportation to low-occupancy driverless cars. For short journeys, compact cities with inviting public spaces and parks would make the daily chore of traveling to work, school, the shops and family a pleasurable part of our day rather than a frustrating time sitting in a vehicle. The design of our cities in both the developed and the developing world will therefore be crucial to locking in low air pollution lifestyles.

We began this chapter with projections for the world's urban population. This is set to double in forty years, but the function and structures of new cities will be determined in the next twenty years.[28] Our current cities need to adapt too. The decades immediately before us will therefore be crucial if we are to reduce the health burden of air pollution and meet our climate change targets.

Above all, we must recognize our air as a precious resource that needs to be protected and not used as a waste disposal mechanism. During our lifetimes we will each breathe around 250 million liters of air, weighing about 300,000 kg.[29] But we do not possess it for more than the few seconds of each breath. It therefore falls to each of us to act where we can, but most of all to governments worldwide to show leadership and to create environments where a low-pollution lifestyle, coupled with low-pollution industry, is no longer a potential goal but the only obvious and rational choice. Reducing the health burden from air pollution is a massive prize for our politicians to seize.

Acknowledgments

My name is on the cover of this book, but I could never have written it without the great team that I have around me.

Huge thanks go to my family and friends, especially to my wife, Cathy, who has supported me through the (mostly) exciting and enjoyable path of writing this book; she has read all of the drafts, made me cups of tea and helped me keep a life balance. I could not have written it without her. I am also grateful to Pamela Davy. I had the pleasure of supervising her PhD and she has read and commented on the book and encouraged me along the way. Thanks also to Becky Fuller for the opening line, which she told me with amazement after a school science lesson, to Judy Martin for reading the early sections, and to Tom and Susan Crossett, who counseled me on the wisdom of writing a book.

Thanks to my wonderful team at King's and also to all the colleagues and fellow scientists who have shared their work and inspired me over the years. Many of you are credited here but, sorry, I could not fit everyone in. Professors Martin Williams, David Fowler and Michal Krzyzanowski told me good stories from their early careers and from working in international organizations. Thanks also to my mum and dad for their tales of the London smogs, and for always being there.

I could not have researched the book without the help of the library services at King's College London. Over the last year I have squeezed in a few hours with my laptop whenever I could. Thank you to the many venues where I sat and wrote: Mike's farm in Pembrokeshire overlooking the Irish Sea, the persistently delayed trains between London and Brighton, Eurostar, TGVs, budget hotels, my mother-in-law Ann's dining room, the many cafés that allowed me to sit in a quiet corner with a pot of tea and also Brighton and Hove's Jubilee Library. The seagulls that I saw from my home-office window and that soared over my head when I worked in the garden were a constant reminder of the invisible air around us.

Finally, I want to thank the team at Melville House, including Steve Gove, and especially my editor Nikki Griffiths who offered me the chance to do this project. Nikki's suggestions and comments have really helped me to define the shape of this book. As scientists we tend to let the results of our published experiments convey what we want to say, but Nikki reminded me that a good book requires stories. In writing this book I have got to know many scientists from the past. I hope that you will enjoy meeting them as much as I have and that this book will encourage you to think about the need to care for the air that we breathe; our ultimate shared resource.

Gary Fuller, Brighton, December 2018

References

Introduction

1. Health Effects Institute and Institute for Health Metrics, *State of Global Air 2017: A special report*. Boston: HEI, 2017.

1 Early explorers

1. Evelyn, J., *Fumifugium, or, The inconveniencie of the aer and smoak of London dissipated together with some remedies humbly proposed*, translated by Gross, A., and Shaw. J. Brighton: Environmental Protection UK, 1661; 2012.
2. Brimblecombe, P., *The Big Smoke*. London: Methuen, 1987; Shaw, N., and Owens, J.S., *The Smoke Problem of Great Cities*. London: Constable & Company, 1925.
3. Thorsheim, P., *Inventing Pollution: Coal, smoke and culture in Britain since 1800*. Athens, Ohio: Ohio University Press, 2006.
4. West, B.J. (2013), "Torricelli and the Ocean of Air: The first measurement of barometric pressure." *Physiology*, Vol. 28, 66–73.
5. Smith, R.A., *Air and Rain: The beginnings of a chemical climatology.* London: Longmans, Green and Company, 1872.
6. Ibid.
7. Clapp, B.W., *An Environmental History of Britain since the Industrial Revolution*. Harlow, Essex: Longman, 1994.
8. Aitken, J. (1888), "On the number of dust particles in the atmosphere." *Nature*, 428–30.

9. Knott, C.G., *Collected Scientific Papers of John Aitken, LL.D., F.R.S., edited for the Royal Society of Edinburgh (with introductory memoir).* s.l.: Cambridge University Press, 1923.
10. Rubin, R.B. (2001), "The history of ozone: The Schönbein period, 1839–1868." *Bulletin for the History of Chemistry*, Vol. 26(1), 40–56.
11. Ibid.
12. Thorsheim, *Inventing Pollution.*
13. Voltz, A., and Kley, D. (1988), "Evaluation of the Montsouris series of ozone measurements." *Nature*, Vol. 332, 240–2.
14. Smith, *Air and Rain.*

2 Warning signs ignored

1. Dr. J.S. Owens, obituary (1942). *Nature*, Vol. 149, 133.
2. Connor, K., "There's something in the air." Wellcome Library, November 20, 2013: http://blog.wellcomelibrary.org/2013/11/theres-something-in-the-air-early-environmental-campaigners/.
3. Owens, J.S. (1936), "Twenty-five years' progress in smoke abatement." *Transactions of the Faraday Society*, Vol. 32, 1234–41.
4. Owens, J.S. (1918), "The measurement of atmospheric pollution." *Quarterly Journal of the Royal Meteorological Society*, Vol. 44, 187.
5. Owens, J.S. (1926), "Measuring the smoke pollution of city air." *The Analyst*, Vol. 51, 2–18.
6. Owens, "Twenty-five years' progress in smoke abatement."
7. Shaw, N., and Owens, J.S., *The Smoke Problem of Great Cities.* London: Constable & Company, 1925.
8. Beaver, H.E.C., "The growth of public opinion," in Mallette, F.S. (ed.), *Problems and Control of Air Pollution.* New York: The American Society of Mechanical Engineers, Reinhold Publishing Corporation, 1955.

9. Owens, J.S. (1922), “Suspended impurity in the air.” *Proceedings of the Royal Society of London*, Series A, Vol. 101(708), 18–37.
10. Shaw and Owens, *The Smoke Problem of Great Cities.*
11. Ibid.
12. Ibid.
13. Whipple, F.J.W. (1929), “Potential gradient and atmospheric pollution: The influence of ‘summer time.’” *Quarterly Journal of the Royal Meteorological Society*, Vol. 55(232), 351–62.
14. Shaw and Owens, *The Smoke Problem of Great Cities.*
15. Taylor, J.S., *Smoke and Health: A lecture delivered at the Manchester College of Technology.* Manchester: Joint Committee of the Manchester and District Smoke Abatement Society and the National Smoke Abatement Society, 1929.
16. Shaw and Owens, *The Smoke Problem of Great Cities.*
17. Ibid.
18. Rollier, A. (1929), “The sun cure and the work cure in surgical tuberculosis.” *British Medical Journal*, Vol. 2(3599), 1206–7.
19. Firket, J. (1936), “Fog along the Meuse Valley.” *Transactions of the Faraday Society*, Vol. 32, 1192–6.
20. Ibid.
21. *Mortality and Morbidity During the London Fog of December 1952.* Reports on public health and medical subjects No. 95. London: Ministry of Health, 1954.
22. McCabe, L.C., and Clayton, G.D. (1952), “Air Pollution by Hydrogen Sulfide in Poza Rica, Mexico. An Evaluation of the Incident of Nov. 24, 1950.” *Archives of Industrial Hygiene and Occupational Medicine*, Vol. 6, 199–213.

3 The great smog

1. The Greater London Authority, *50 Years On: The struggle for air quality in London since the great smog of December 1952*. London: The Greater London Authority, 2002; Brimblecombe, P., *The Big Smoke*. London: Methuen, 1987.
2. Ministry of Health, *Mortality and Morbidity During the London Fog of December 1952*. Reports on public health and medical subjects No. 95. London: 1954.
3. Ibid.
4. Ibid.
5. Wilkins, E.T. (1954), "Air pollution and the London fog of December, 1952." *Journal of the Royal Sanitary Institute*, Vol. 74(1), 1–21.
6. Logan, W.P.D. (1953), "Mortality in the London fog incident, 1952." *The Lancet*, Vol. 261(6755), 226–338; Wilkins, E.T. (1954), "Air pollution aspects of the London fog of 1952." *Journal of the Royal Meteorological Society*, 267–71.
7. Thorsheim, P., *Inventing Pollution: Coal, smoke and culture in Britain since 1800*. Athens, Ohio: Ohio University Press, 2006.
8. Ibid.
9. Bell, M.L., and Davis, D.L. (2001), "Reassessment of the lethal London fog of 1952: Novel indicators of acute and chronic consequences of acute exposure to air pollution." *Environmental Health Perspectives*, Vol. 109, Supplement 3, 389–94.
10. Ibid.
11. Clapp, B.W., *An Environmental History of Britain since the Industrial Revolution*. Harlow, Essex: Longman, 1994.
12. Hansard, February 2, 1953. Vols. 510, cc1460–62.
13. Hansard, 1953. Vols. 515, cc841–52.

14. Sir Hugh Eyre Campbell Beaver KBE LLD, obituary. The Institution of Civil Engineers, 1967.
15. Cavendish, R., "Publication of the Guinness Book of Records." *History Today*: https://www.questia.com/magazine/1G1-135180380/publication-of-the-guinness-book-of-records-august.
16. Beaver, H.E.C., "The growth of public opinion," in Mallette, F.S. (ed.), *Problems and Control of Air Pollution.* New York: The American Society of Mechanical Engineers, Reinhold Publishing Corporation, 1955; Wilkins, "Air pollution and the London fog of December, 1952."
17. Thorsheim, *Inventing Pollution.*
18. Clapp, *An Environmental History of Britain.*
19. Ibid.
20. Sutherland, J., "Sir Gerald and the roundabout." *The Guardian*, December 27, 1999: https://www.theguardian.com/uk/1999/dec/27/hamiltonvalfayed.features11; "Gerald Nabarro." Wikipedia: https://en.wikipedia.org/wiki/Gerald_Nabarro.
21. Clapp, *An Environmental History of Britain.*
22. Ministry of Health, *Mortality and Morbidity During the London Fog of December 1952*; Logan, W.P.D. (1956), "Mortality from fog in London, January, 1956." *British Medical Journal*, Vol. 1(4969), 722; Brimblecombe, *The Big Smoke*; Anderson, H.R., Limb, E.S., Bland, J.M., De Leon, A.P., Strachan, D.P., and Bower, J.S. (1995), "Health effects of an air pollution episode in London, December 1991." *Thorax*, Vol. 50(11), 1188–93; Stedman, J.R. (2004), "The predicted number of air pollution related deaths in the UK during the August 2003 heatwave." *Atmospheric Environment*, Vol. 38(8), 1087–90; Macintyre, H.L., Heaviside, C., Neal, L.S., Agnew, P., Thornes, J., and Vardoulakis, S. (2016), "Mortality and emergency hospitalizations

associated with atmospheric particulate matter episodes across the UK in spring 2014." *Environment International*, Vol. 97, 108–16.

23. Anderson et al., "Health effects of an air pollution episode in London, December 1991."

4 The madness of lead in petrol

1. Bess, M. (2002), review of McNeill, J.R., *Something New Under the Sun: An Environmental History of the Twentieth-Century World* (New York: W.W. Norton, 2001). *Journal of Political Ecology*. Volume 9, 1-2.
2. Pearce, F., "Inventor hero was a one-man environmental disaster." *New Scientist,* June 7, 2017.
3. Grandjean, P., Bailar, J.C., Gee, D., Needleman, H.L., Ozonoff, D.M., Richter, E., Sofritti, M., and Soskolne, C.L. (2003), "Implications of the precautionary principle in research and policy-making." *American Journal of Industrial Medicine*, Vol. 45(4), 382–5.
4. Ibid.
5. Tilton, G., *Clair Cameron Patterson, 1922–1995: A Biographical Memoir*. Washington, DC: National Academy of Sciences, 1998.
6. Needleman, H.L., Gunnoe, C., Leviton, A., Reed, R., Peresie, H., Maher, C., and Barrett, P. (1979), "Deficits in psychologic and classroom performance of children with elevated dentine lead levels." *New England Journal of Medicine*, Vol. 300(13), 689–95; Carey, B., "Dr. Herbert Needleman, Who Saw Lead's Wider Harm to Children, Dies at 89." *New York Times*, July 27, 2017.
7. Carey, "Dr. Herbert Needleman."
8. Grandjean et al., "Implications of the precautionary principle."
9. Chesshyre, R., "Des Wilson: 'We can only try to edge the world in the right direction.'" *The Independent*, February 28, 2011.

10. Leigh, D., Evans, R., and Mahmood, M., "Killer chemicals and greased palms—the deadly 'end game' for leaded petrol." *The Guardian*, June 30, 2010.
11. Lanphear, B.P., Rauch, S., Auinger, P., Allen, R.W., and Hornung, R.W. (2018), "Low-level lead exposure and mortality in US adults: A population-based cohort study." *The Lancet Public Health*, Vol. 3(4), 177–84.

5 Ozone, the pollutant that rots rubber

1. South Coast Air Quality Management District (1997), "The Southland's War on Smog: Fifty Years of Progress Toward Clean Air (through May 1997)," http://www.aqmd.gov/home/research /publications/50-years-of-progress.
2. Dunsey, J., "Localising smog—transgressions in the therapeutic landscape," in DuPuis, E.M. (ed.), *Smoke and Mirrors: The politics and culture of air pollution*. New York: New York University Press, 2004.
3. Cohen, S. K., Interview with Zus (Maria) Haagen-Smit (1910–2006). Pasadena, California: Archives of the California Institute of Technology, 2000.
4. Haagen-Smit, A.J. (1952), "Chemistry and physiology of Los Angeles smog." *Industrial and Engineering Chemistry*, Vol. 44(6), 1342–6.
5. Cohen, Interview with Zus (Maria) Haagen-Smit.
6. Kean, S. (2016), "The flavor of smog." *Distillations Magazine, The Science History Institute*. https://www.sciencehistory.org/distillations /magazine/the-flavor-of-smog.
7. Royal College of Physicians, *Air Pollution and Health*. London: Pitman, 1970.
8. Atkins, D.H.F., Cox, R.A., and Eggleton, A.E.J. (1972),

"Photochemical ozone and sulphuric acid aerosol formation in the atmosphere over southern England." *Nature*, Vol. 235(5338), 372–6.

9. Jones, T., Overy, C., and Tansey, E.M., *Air Pollution Research in Britain c1955–c2000*. London: The Wellcome Trust, 2016.
10. Cox, R.A., Eggleton, A.E.J., Derwent, R.G., Lovelock, J.E., and Pack, D.H. (1975), "Long-range transport of photochemical ozone in North-Western Europe." *Nature*, Vol. 255(5504), 118–21.
11. Jones et al., *Air Pollution Research in Britain.*
12. Jenkin, M.E., Davies, T.J., and Stedman, J.R. (2002), "The origin and day-of-week dependence of photochemical ozone episodes in the UK." *Atmospheric Environment*, Vol. 36(6), 999–1012.
13. Stedman, J.R. (2004), "The predicted number of air pollution related deaths in the UK during the August 2003 heatwave." *Atmospheric Environment*, Vol. 38(8), 1087–90.
14. World Health Organization, Regional Office for Europe, *Review of Evidence on the Health Aspects of Air Pollution—REVIHAAP Project, technical report.* Bonn: WHO, 2013.
15. Di, Q., Wang, Y., Zanobetti, A., Wang, Y., Koutrakis, P., Choirat, C., Dominici, F., and Schwartz, J.D. (2017), "Air pollution and mortality in the Medicare population." *New England Journal of Medicine*, Vol. 376(26), 2513–22.

6 Acid rain and the particles that form in our air

1. United Nations, *Clearing the Air: 25 years of the Convention on Long-Range Transboundary Air Pollution*. Geneva and New York: United Nations, 2004.
2. Clapp, B.W., *An Environmental History of Britain since the Industrial Revolution*. Harlow, Essex: Longman, 1994.

3. Ottar, B. (1976), "Organization of long range transport of air pollution monitoring in Europe." *Water, Air, and Soil Pollution*, Vol. 6, 219–29.
4. Hollingshead, I., "Whatever happened to the acid rain debate?" *The Guardian*, October 21, 2005: https://www.theguardian.com/news/2005/oct/22/mainsection.saturday32.
5. Ottar, B. (1977), "International agreement needed to reduce long-range transport of air pollutants in Europe." *Ambio*, Vol. 6(5), 262–9.
6. United Nations, *Clearing the Air*.
7. Ottar, "International agreement needed."
8. Ottar, "Organization of long range transport."
9. Clapp, *An Environmental History of Britain*; Rees, R.L., "Removal of sulfur dioxide from power plant stack gases," in Mallette, F.S. (ed.), *Problems and Control of Air Pollution*. New York: The American Society of Mechanical Engineers, Reinhold Publishing Corporation, 1955.
10. Barns, R. A. (1977), "Sulphur deposit account." *Nature*, Vol. 268, 92–3.
11. Editorial (July 14, 1977), "Million dollar problem—billion dollar solution?" *Nature*, Vol. 268, 89.
12. Barnes, R., Parkinson, G.S., and Smith, A.E. (1983), "The costs and benefits of sulphur oxide control." *Journal of the Air Pollution Control Association*, Vol. 33(8), 737–41.
13. Ball, D.J. and Hume, R. (1977), "The relative importance of vehicular and domestic emissions of dark smoke in Greater London in the mid-1970s, the significance of smoke shade measurements, and an explanation of the relationship of smoke shade to gravimetric." *Atmospheric Environment*, Vol. 11(11), 1065–73.
14. Ball, D.J. (1977), "Sampling. Some measurements of atmospheric pollution by aerosols in an urban environment." *Proceedings of the Analytical Division of the Chemical Society*, Vol. 14(8), 203–8.

15. Expert Panel on Air Quality Standards, *Particles*. London: Department of Environment, Transport and the Regions, 1998.
16. Stedman, J. (1997), "A UK wide episode of elevated particle (PM10) concentration in March 1996." *Atmospheric Environment*, Vol. 31(15), 2381–3.
17. Ibid.
18. Macintyre, H.L., Heaviside, C., Neal, L.S., Agnew, P., Thornes, J., and Vardoulakis, S. (2016), "Mortality and emergency hospitalizations associated with atmospheric particulate matter episodes across the UK in spring 2014." *Environment International*, Vol. 97, 108–16.
19. European Environment Agency, *Air Quality in Europe—2016 report*. EA Report number 28/2016. Luxembourg: EEA, 2016.
20. Wang, S. and Hao, J. (2012), "Air quality management in China: Issues, challenges, and options." *Journal of Environmental Sciences*, Vol. 24(1), 2–13.
21. Turnock, S.T., Butt, E.W., Richardson, T.B., Mann, G.W., Reddington, C.L., Forster, P.M., Haywood, J., Crippa, M., Janssens-Maenhout, G., Johnson, C.E., and Bellouin, N. (2016), "The impact of European legislative and technology measures to reduce air pollutants on air quality, human health and climate." *Environmental Research Letters*, Vol. 11(2), 024010.

7 A tale of six cities

1. Dockery, D.W., Pope, C.A., Xu, X., Spengler, J.D., Ware, J.H., Fay, M.E., Ferris, B.G., Jr., and Speizer, F.E. (1993), "An association between air pollution and mortality in six US cities." *New England Journal of Medicine*, Vol. 329(24), 1753–9.
2. Ibid.

3. The Health Effects Institute, *Reanalysis of the Harvard Six Cities Study and the American Cancer Society Study of Particulate Mortality: A special report of the Institute's particle epidemiology reanalysis project.* Cambridge, MA: HEI, 2000.
4. Moolgavkar, S.H., and Luebeck, E.G. (1996), "A critical review of the evidence on particulate air pollution and mortality." *Epidemiology*, Vol. 7(9), 420–8.
5. Vedal, S. (1997), "Ambient particles and health: Lines that divide." *Journal of the Air & Waste Management Association*, Vol. 47(5), 551–81.
6. Health Effects Institute, *Reanalysis of the Harvard Six Cities Study.*
7. Laden, F., Schwartz, J., Speizer, F.E., and Dockery, D.W. (2006), "Reduction in fine particulate air pollution and mortality: Extended follow-up of the Harvard Six Cities study." *American Journal of Respiratory and Critical Care Medicine*, Vol. 173(6), 667–72.
8. Gauderman, W.J., McConnell, R., Gilliland, F., London, S., Thomas, D., and Avol, E. (2000), "Association between air pollution and lung function growth in southern California children." *American Journal of Respiratory Critical Care Medicine*, Vol. 162(4), 1383–90.
9. Gauderman, W.J., Urman, R., Avol, E., Berhane, K., McConnell, R., Rappaport, E., Chang, R., Lurmann, F., and Gilliland, F. (2015), "Association of improved air quality with lung development in children." *New England Journal of Medicine*, Vol. 372(10), 905–13.
10. Ministry of Health, *Mortality and Morbidity During the London Fog of December 1952.* Reports on public health and medical subjects No. 95. London: 1954.
11. Royal College of Physicians and Royal College of Paediatrics and Child Health, *Every Breath We Take: The lifelong impact of air pollution*. London: Royal College of Physicians, 2016.

12. Black, D., "Sellafield: the nuclear legacy." *The New Scientist*, March 7, 1985.
13. Hansell, A., Ghosh, R.E., Blangiardo, M., Perkins, C., Vienneau, D., Goffe, K., Briggs, D., and Gulliver, J. (2016), "Historic air pollution exposure and long-term mortality risks in England and Wales: Prospective longitudinal cohort study." *Thorax*, Vol. 71(4), 330–8.
14. Kelly, F.J. (2003), "Oxidative stress: Its role in air pollution and adverse health effects." *Occupational and Environmental Medicine*, Vol. 60(8), 612–6.
15. Pirani, M., Best, N., Blangiardo, M., Liverani, S., Atkinson, R.W., and Fuller, G.W. (2015), "Analysing the health effects of simultaneous exposure to physical and chemical properties of airborne particles." *Environment International*, Vol. 79, 56–64.
16. Health Effects Institute and Institute for Health Metrics, *State of Global Air 2017: A special report*. Boston: HEI, 2017.
17. Di, Q., Wang, Y., Zanobetti, A., Wang, Y., Koutrakis, P., Choirat, C., Dominici, F., and Schwartz, J.D. (2017), "Air pollution and mortality in the Medicare population." Vol. 376(26), 2513–22.
18. Health Effects Institute, *State of Global Air 2017*.

8 A global tour in a polluted world

1. Transport and Environment, *Diesel: The true and dirty story*. Brussels: T&E, 2017.
2. European Environment Agency, *Air Quality in Europe—2016 report*, EA Report number 28/2016. Luxembourg: EEA, 2016.
3. Goudie, S. (2014), "Desert dust and human health disorders." *Environment International*, Vol. 63, 101–13.
4. Uno, I., Eguchi, K., Yumimoto, K., Takemura, T., Shimizu, A.,

Uematsu, M., Liu, Z., Wang, Z., Hara, Y., and Sugimoto, N. (2009), "Asian dust transported one full circuit around the globe." *Nature Geoscience*, Vol. 2(8).

5. Health Effects Institute and Institute for Health Metrics, *State of Global Air 2017: A special report*. Boston: HEI, 2017.
6. Johnston, F.H., Henderson, S., Chen, Y., Randerson, J.Y., Marlier, M., DeFries, R.S., Kinney, P., Bowman, D.J.M.S., and Brauer, M. (2012), "Estimated global mortality attributable to smoke from landscape fires." *Environmental Health Perspectives*, Vol. 120(5), 695.
7. Tian, L., Ho, K., Louie, P.K.K., Qiu, H., Pun, V.C., Kan, H., Ignatius, T.S., and Wong, T.W. (2013), "Shipping emissions associated with increased cardiovascular hospitalizations." *Atmospheric Environment*, Vol. 74, 320–5.
8. Schmidt, A., Ostro, B., Carslaw, K.S., Wilson, M., Thordarson, T., Mann, G.W., and Simmons, A.J. (2011), "Excess mortality in Europe following a future Laki-style Icelandic eruption." *Proceedings of the National Academy of Sciences*, Vol. 108(38), 15710–5.
9. Helmig, D., Rossabi, S., Hueber, J., Tans, P., Montzka, S.A., Masarie, K., Thoning, K., Plass-Duelmer, C., Claude, A., Carpenter, L.J., and Lewis, A. (2016), "Reversal of global atmospheric ethane and propane trends largely due to US oil and natural gas production." *Nature Geoscience*, 490–5.
10. Roberts, D., "Opinion: How the US embassy Tweeted to clear Beijing's air." *Wired*, June 3, 2015.
11. Anejionu, O.C., Whyatt, J.D., Blackburn, G.A., and Price, C.S. (2015), "Contributions of gas flaring to a global air pollution hotspot: Spatial and temporal variations, impacts and allevation." *Atmospheric Environment*, Vol. 118, 184–93.

12. Miller, J., and Façanha, C., *The State of Clean Transport Policy*. Washington: ICCT, 2014.
13. Broome, R.A., Fann, N., Cristina, T.J.N., Fulcher, C., Duc, H., and Morgan, G.G. (2015), "The health benefits of reducing air pollution in Sydney, Australia." *Environmental Research,* Vol. 143, 19–25.
14. Fuller, G., "All is not pristine in New Zealand's polluted air." *The Guardian*, August 28, 2016: https://www.theguardian.com/environment/2016/aug/28/pollution-new-zealand-wood-fires-insulation-world-weatherwatch.
15. Clean Air Institute, *Air Quality in Latin America*. Washington: Clean Air Institute, 2013.
16. de Fatima Andrade, M., Kumar, P., de Freitas, E.D., Ynoue, R.Y., Martins, J., Martins, L.D., Nogueira, T., Perez-Martinez, P., de Miranda, R.M., Albuquerque, T., and Gonçalves, F.L.T. (2017), "Air quality in the megacity of São Paulo: Evolution over the last 30 years and future perspectives." *Atmospheric Environment*, Vol. 159, 66–82.
17. Roberts, "Opinion: How the US embassy Tweeted to clear Beijing's air."
18. Chai, F., Gao, J., Chen, Z., Wang, S., Zhang, Y., Zhang, J., Zhang, H., Yun, Y., and Ren, C. (2015), "Spatial and temporal variation of particulate matter and gaseous pollutants in 26 cities in China." *Journal of Environmental Sciences*, Vol. 26(1), 75–82.
19. Wong, E., "China lets media report on air pollution crisis." *New York Times*, January 14, 2013.
20. Song, C., He, J., Wu, L., Jin, T., Chen, X., Li, R., Ren, P., Zhang, L., and Mao, H. (2017), "Health burden attributable to ambient PM2.5 in China." *Environmental Pollution*, Vol. 223, 575–86.
21. Ebenstein, A., Fan, M., Greenstone, M., He, G., and Zhou, M. (2017), "New evidence on the impact of sustained exposure to air pollution

on life expectancy from China's Huai River Policy." *Proceedings of the National Academy of Sciences*, Vol. 114(39), 10384–8.

22. Shaddick, G., Thomas, M.L., Green, A., Brauer, M., Donkelaar, A., Burnett, R., Chang, H.H., Cohen, A., Dingenen, R.V., Dora, C., and Gumy, S. (2017), "Data integration model for air quality: A hierarchical approach to the global estimation of exposures to ambient air pollution." *Journal of the Royal Statistical Society*, Series C (Applied Statistics), Vol. 67(1), 231–53.
23. Health Effects Institute and the Institute for Health Metrics, *The State of Global Air—2018*. Boston: HEI, 2018.
24. Health Effects Institute, *State of Global Air 2017*.
25. Royal Society, *Ground-level Ozone in the 21st Century: Future trends, impacts and policy implications*. London: Royal Society, 2008; Van Dingenen, R., Dentener, F.J., Raes, F., Krol, M.C., Emberson, L., and Cofala, J. (2009), "The global impact of ozone on agricultural crop yields under current and future air quality legislation." *Atmospheric Environment*, Vol. 43(3), 604–18.
26. Monks, P. (2000), "A review of the observations and origins of the spring ozone maximum." *Atmospheric Environment*, Vol. 34(21), 3545–61; Royal Society, *Ground-level Ozone in the 21st Century.*
27. Royal Society, *Ground-level Ozone in the 21st Century*.
28. McDonald, B.C., de Gouw, J.A., Gilman, J.B., Jathar, S.H., Akherati, A., Cappa, C.D., Jimenez, J.L., Lee-Taylor, J., Hayes, P.L., McKeen, S.A., and Cui, Y.Y. (2018), "Volatile chemical products emerging as largest petrochemical source of urban organic emissions." *Science*, Vol. 359(6377), 760–4.
29. Ahmadov, R., McKeen, S., Trainer, M., Banta, R., Brewer, A., Brown, S., Edwards, P.M., de Gouw, J.A., Frost, G.J., Gilman, J., and Helmig,

D. (2015), "Understanding high wintertime ozone pollution events in an oil- and natural gas-producing region of the western US." *Atmospheric Physics and Chemistry*, Vol. 15(1), 411–29.

30. Peischl, J., Ryerson, T.B., Aikin, K.C., Gouw, J.A., Gilman, J.B., Holloway, J.S., Lerner, B.M., Nadkarni, R., Neuman, J.A., Nowak, J.B., and Trainer, M. (2014), "Quantifying atmospheric methane emissions from the Haynesville, Fayetteville, and northeastern Marcellus shale gas production regions." *Journal of Geophysical Research: Atmospheres*, Vol. 120(5), 2119–39.
31. Franco, B., Bader, W., Toon, G.C., Bray, C., Perrin, A., Fischer, E.V., Sudo, K., Boone, C.D., Bovy, B., Lejeune, B., and Servais, C. (2015), "Retrieval of ethane from ground-based FTIR solar spectra using improved spectroscopy: Recent burden increase above Jungfraujoch." *Journal of Quantitative Spectroscopy & Radiative Transfer*, Vol. 160, 36–49.
32. Helmig, D., Rossabi, S., Hueber, J., Tans, P., Montzka, S.A., Masarie, K., Thoning, K., Plass-Duelmer, C., Claude, A., Carpenter, L.J., and Lewis, A. (2016), "Reversal of global atmospheric ethane and propane trends largely due to US oil and natural gas production." *Nature Geoscience*, Vol. 9, 490–5.
33. Roohani, Y.H., Roy, A.A., Heo, J., Robinson, A.L., and Adams, P.J. (2017), "Impact of natural gas development in the Marcellus and Utica shales on regional ozone and fine particulate matter levels." *Atmospheric Environment*, Vol. 155, 11–20.
34. Inman, M. (2016), "Can fracking power Europe?" *Nature News*, Vol. 531, 22–4.
35. Alvarez, R.A., Pacala, S.W., Winebrake, J.J., Chameides, W.L., and Hamburg, S.P. (2012), "Greater focus needed on methane leakage from natural gas infrastructure." *Proceedings of the National Academy*

of Sciences, Vol. 109(17); Peischl et al., "Quantifying atmospheric methane emissions."

36. World Health Organization, *Global Urban Ambient Air Pollution Database (update 2016).* Geneva: World Health Organization, 2016.

9 Counting particles and the enigma of modern air pollution

1. Seaton, A., Godden, D., MacNee, W., and Donaldson, K. (1995), "Particulate air pollution and acute health effects." *The Lancet*, Vol. 345(8943), 176–8; Seaton, A. (1996), "Particles in the air: The enigma of urban air pollution." Journal of the Royal Society of Medicine, Vol. 89(11), 604–7.
2. Anderson, H.R., Limb, E.S., Bland, J. M., De Leon, A.P., Strachan, D.P., and Bower, J.S. (1995), "Health effects of an air pollution episode in London, December 1991." *Thorax*, Vol. 50(11), 1188–93.
3. "About." The Institute of Occupational Medicine: http://www.iomworld.org/about/.
4. The Royal Society and the Royal Academy of Engineering. *Nanoscience and Nanotechnologies*. London: 2005.
5. Atkinson, R.W., Fuller, G.W., Anderson, H.R., Harrison, R.M., and Armstrong, B. (2010), "Urban ambient particle metrics and health: A time-series analysis." *Epidemiology*, Vol. 21(4), 501–11.
6. Jones, A.M., Harrison, R.M., Barratt, B., and Fuller, G. (2012), "A large reduction in airborne particle number concentrations at the time of the introduction of 'sulphur free' diesel and the London low emission zone." *Atmospheric Environment*, Vol. 50, 129–38.
7. Hudda, N., and Fruin, S.A. (2016), "International airport impacts to air quality: Size and related properties of large increases in

ultrafine particle number concentrations." *Environmental Science & Technology*, Vol. 50(7), 3362–70.

8. Keuken, M.P., Moerman, M., Zandveld, P., Henzing, J.S., and Hoek, G. (2015), "Total and size-resolved particle number and black carbon concentrations in urban areas near Schiphol airport (the Netherlands)." *Atmospheric Environment*, Vol. 104, 132–42.
9. Hansell, A.L., Blangiardo, M., Fortunato, L., Floud, S., de Hoogh, K., Fecht, D., Ghosh, R.E., Laszlo, H.E., Pearson, C., Beale, L., and Beevers, S. (2013), "Aircraft noise and cardiovascular disease near Heathrow airport in London: Small area study." *British Medical Journal*, Vol. 34, f5432.
10. Barrett, S.R., Yim, S.H., Gilmore, C.K., Murray, L.T., Kuhn, S.R., Tai, A.P., Yantosca, R.M., Byun, D.W., Ngan, F., Li, X., and Levy, J.I. (2012), "Public health, climate, and economic impacts of desulfurizing jet fuel." *Environmental Science & Technology*, Vol. 46, 4275–82.
11. Abernethy, R.C., Allen, R.W., McKendry, I.G., and Brauer, M. (2013), "A land use regression model for ultrafine particles in Vancouver, Canada." *Environmental Science & Technology*, Vol. 47(10), 5217–25.
12. Vert, C., Meliefste, K., and Hoek, G. (2016), "Outdoor ultrafine particle concentrations in front of fast food restaurants." *Journal of Exposure Science and Environmental Epidemiology*, Vol. 26(1), 35.
13. Brines, M., Dall'Osto, M., Beddows, D.C.S., Harrison, R.M., Gómez-Moreno, F., Núñez, L., Artíñano, B., Costabile, F., Gobbi, G.P., Salimi, F., and Morawska, L. (2015), "Traffic and nucleation events as main sources of ultrafine particles in high-insolation insolation developed world cities." *Atmospheric Chemistry and Physics*, 5929–45.
14. Beddows, D.C.S., Harrison, R.M., Green, D.C., and Fuller, G.W. (2015), "Receptor modelling of both particle composition and size

distribution from a background site in London, UK." *Atmospheric Chemistry and Physics*, Vol. 15(17), 10107–25.

10 VW and the tricky problem with diesel

1. Transport and Environment, *Diesel: The true and dirty story*. Brussels: T&E, 2017.
2. Ibid.
3. European Parliament Committee of Inquiry into Emission Measurements in the Automotive Sector, *Report on the Inquiry into Emission Measurements in the Automotive Sector (2016/2215(INI))*. European Parliament, 2016.
4. Transport and Environment, *Diesel*.
5. Cames, M., and Helmers, E. (2013), "Critical evaluation of the European diesel car boom-global comparison, environmental effects and various national strategies." *Environmental Sciences Europe*, Vol. 25(1), 15.
6. Ibid.
7. Transport and Environment, *Diesel*.
8. Carslaw, D.C., Beevers, S.D., and Fuller, G. (2001), "An empirical approach for the prediction of annual mean nitrogen dioxide concentrations in London." *Atmospheric Environment*, Vol. 35(8), 1505–15.
9. Carslaw, D.C. (2005), "Evidence of an increasing NO2/NOX emissions ratio from road traffic emissions." *Atmospheric Environment*, Vol. 39(26).
10. Font, A., Guiseppin, L., Ghersi, V., and Fuller, G.W. (2018), "A tale of two cities: Is air pollution improving in London and Paris?" Not yet published.

11. Department for Environment, Food and Rural Affairs, *Valuing Impacts on Air Quality: Updates in valuing changes in emissions of oxides of nitrogen (NOX) and concentrations of nitrogen dixoide (NO2)*. London: Defra, 2015.
12. Carslaw, D.C., and Rhys-Tyler, G. (2013), "New insights from comprehensive on-road measurements of NOx, NO2 and NH3 from vehicle emission remote sensing in London, UK." *Atmospheric Environment*, Vol. 81, 339–47.
13. Lichfield, J., "The 2CV - A French icon: La toute petite voiture." *The Independent*, April 18, 2008: http://www.independent.co.uk/news/world/europe/the-2cv-a-french-icon-la-toute-petite-voiture-811246.html.
14. Department for Transport, *Vehicle Emissions Testing Programme*. London: DfT, 2016.
15. Ibid.
16. Hagman, R., Weber, C., and Amundsen, A.H., *Emissions from New Vehicles—Trustworthy?* (English summary). Oslo: TOI, 2015.
17. Sjödin, Å., Jerksjö, M., Fallgren, H., Salberg, H., Parsmo, R., and Hult, C., *On-Road Emission Performance of Late Model Diesel and Gasoline Vehicles as Measured by Remote Sensing*. Stockholm: IVL, 2017.
18. Department for Transport, *Vehicle Emissions Testing Programme*.
19. German, J., "The emissions test defeat device problem in Europe is not about VW." International Council for Clean Transport, May 12, 2016: http://www.theicct.org/blogs/staff/emissions-test-defeat-device-problem-europe-not-about-vw.
20. Font et al., "A tale of two cities."
21. Ibid.
22. Grange, S.K., Lewis, A.C., Moller, S.J., Carslaw, D.C. (2017), "Lower vehicular primary emissions of NO2 in Europe than assumed in

policy projections." *Nature Geosciences*, Vol. 10, 914–18; Sjödin et al., *On-road emission performance.*

23. Carslaw, "Evidence of an increasing NO2/NOX emissions ratio from road traffic emissions."
24. Font, A., and Fuller, G.W. (2016), "Did policies to abate atmospheric emissions from traffic have a positive effect in London?" *Environmental Pollution*, Vol. 218, 463–74.
25. Sjödin et al., *On-Road Emission Performance.*
26. Dunmore, R.E., Hopkins, J.R., Lidster, R.T., Lee, J.D., Evans, M.J., Rickard, A.R., Lewis, A.C., and Hamilton, J.F. (2015), "Diesel-related hydrocarbons can dominate gas phase reactive carbon in megacities." *Atmospheric Chemistry and Physics*, Vol. 15(17), 9983–96.
27. Ibid.

11 Wood-burning—the most natural way to heat my home?

1. Favez, O., Cachier, H., Sciare, J., Sarda-Estève, R., and Martinon, L. (2009), "Evidence for a significant contribution of wood-burning aerosols to PM 2.5 during the winter season in Paris, France." *Atmospheric Environment*, Vol. 43(22), 3640–4.
2. Wagener, S., Langner, M., Hansen, U., Moriske, H.J., and Endlicher, W.R. (2012), "Spatial and seasonal variations of biogenic tracer compounds in ambient PM 10 and PM 1 samples in Berlin, Germany." *Atmospheric Environment*, Vol. 47, 33–42.
3. Fuller, G.W., Sciare, J., Lutz, M., Moukhtar, S., and Wagener, S. (2013), "New directions: Time to tackle urban wood-burning?" *Atmospheric Environment*, Vol. 68, 295–6.
4. Fuller, G.W., Tremper, A.H., Baker, T.D., Yttri, K.E., and Butterfield,

D. (2014), "Contribution of wood-burning to PM10 in London." *Atmospheric Environment*, Vol. 87, 87–94.

5. Fuller et al., "New directions."
6. Walters, E., *Summary Results of the Domestic Wood Use Survey.* London: Department for Energy and Climate Change, 2016.
7. Font, A., and Fuller, G.W., *Airborne Particles from Wood-Burning in UK Cities.* London: King's College London, 2017.
8. Reis, F., Marshall, J.D., and Brauer, M. (2009), "Intake fraction of urban wood smoke." *Atmospheric Environment*, 4701–6.
9. Mullholland, R., "Segolene Royal defeats 'ridiculous' Paris ban on open log fires." *Daily Telegraph*, December 30, 2014: http://www.telegraph.co.uk/news/worldnews/europe/france/11317811/Segolene-Royal-defeats-ridiculous-Paris-ban-on-open-log-fires.html.
10. Petersen, L.K. (2008), "Autonomy and proximity in household heating practices: The case of wood-burning stoves." *Journal of Environmental Policy and Planning*, Vol. 10(4), 423–38.
11. Robinson, D.L. (2016), "What makes a successful woodsmoke-reduction program?" *Air Quality and Climate Change*, Vol. 50(3), 25–33.
12. Wright, T., "Special report: how polluted are New Zealand's rivers?" Newshub, February 27, 2018, http://www.newshub.co.nz/home/new-zealand/2017/02/special-report-how-polluted-are-new-zealand-s-rivers.html.
13. Coulson, G., Bian, R., and Somervell, E. (2015), "An investigation of the variability of particulate emissions from woodstoves in New Zealand." *Aerosol and Air Quality Research*, Vol. 15, 2346–56.
14. Cupples, J., Guyatt, V., and Pearce, J. (2007), "'Put on a jacket, you wuss': Cultural identities, home heating, and air pollution in Christchurch, New Zealand." *Environment and Planning A*, Vol. 39(12), 2883–98.

15. Valiente, G., "New rules for wood-burning appliances in Montreal, two decades after ice storm." *The Globe and Mail*, January 4, 2018.
16. Whitehouse, A.C., Black, C.B., Heppe, M.S., Ruckdeschel, J., and Levin, S.M. (2008), "Environmental exposure to Libby asbestos and mesotheliomas." *American Journal of Industrial Medicine*, Vol. 51(11), 877–80.
17. Noonan, C.W., Navidi, W., Sheppard, L., Palmer, C.P., Bergauff, M., Hooper, K., and Ward, T.J. (2012), "Residential indoor PM2. 5 in wood stove homes: Follow-up of the Libby changeout program." *Indoor Air*, Vol. 22(6), 492–500.
18. Noonan, C.W., Ward, T.J., Navidi, W., and Sheppard, L. (2012), "A rural community intervention targeting biomass combustion sources: Effects on air quality and reporting of children's respiratory outcomes." *Occupational and Environmental Medicine*, Vol. 69 (5), 354–60.
19. Coulson et al., "An investigation of the variability of particulate emissions from woodstoves in New Zealand."
20. Yap, P.S., and Garcia, C. (2015), "Effectiveness of residential wood-burning regulation on decreasing particulate matter levels and hospitalizations in the San Joaquin Valley air basin." *American Journal of Public Health*, Vol. 105(4), 772–8.
21. Johnston, F.H., Hanigan, I.C., Henderson, S.B., and Morgan, G.G. (2013), "Evaluation of interventions to reduce air pollution from biomass smoke on mortality in Launceston, Australia: Retrospective analysis of daily mortality, 1994–2007." *British Medical Journal*, Vol. 346, e8446.
22. Robinson, "What makes a successful woodsmoke-reduction program?"

23. Davy, P.K., Ancelet, T., Trompetter, W.J., Markwitz, A., and Weatherburn, D.C. (2012), "Composition and source contributions of air particulate matter pollution in a New Zealand suburban town." *Atmospheric Pollution Research,* Vol. 3(1), 143–7.
24. Cavanagh, J.E., Davy, P., Ancelet, T., and Wilton, E. (2012), "Beyond PM10: benzo(a)pyrene and As concentrations in New Zealand air." *Air Quality and Climate Change,* Vol. 46(2), 15.
25. Phaphitis, N., "In crisis, Greeks turn to wood-burning—and choke." *Ekathimerini,* January 1, 2013: http://www.ekathimerini.com/147932/article/ekathimerini/community/in-crisis-greeks-turn-to-wood-burning-and-choke.
26. Airuse, *Biomass Burning in Southern Europe.* Barcelona: Airuse Project, 2015.
27. Health Effects Institute and Institute for Health Metrics, *State of Global Air 2017: A special report.* Boston: HEI, 2017; Landrigan, P., et al., *The Lancet Commission on Pollution and Health.* The Lancet, 2017.
28. Bruns, E.A., Krapf, M., Orasche, J., Huang, Y., Zimmermann, R., Drinovec, L., Močnik, G., El-Haddad, I., Slowik, J.G., Dommen, J., and Baltensperger, U. (2015), "Characterization of primary and secondary wood combustion products generated under different burner loads." *Atmospheric Chemistry and Physics,* Vol. 15(5), 2825–41.
29. Williams, M.L., Lott, M.C., Kitwiroon, N., Dajnak, D., Walton, H., Holland, M., Pye, S., Fecht, D., Toledano, M.B., and Beevers, S.D. (2018), "*The Lancet* countdown on health benefits from the UK Climate Change Act: A modelling study for Great Britain." *The Lancet Planetary Health,* Vol. 2(5), 205–13.
30. Brack, D., *Woody Biomass for Power and Heat: Impacts on the global climate.* London: Chatham House, The Royal Institute for

International Affairs, 2017; Laganière, J., Paré, D., Thiffault, E., and Bernier, P.Y. (2016), "Range and uncertainties in estimating delays in greenhouse gas mitigation potential of forest bioenergy sourced from Canadian forests." *GCB Bioenergy*, Vol. 9(2), 358–69.

31. Bølling, A.K., Pagels, J., Yttri, K.E., Barregard, L., Sallsten, G., Schwarze, P.E., and Boman, C. (2009), "Health effects of residential wood smoke particles: The importance of combustion conditions and physicochemical particle properties." *Particle and Fibre Toxicology*, Vol. 61(1), 65.
32. Air Quality Expert Group, *The Potential Air Quality Impacts from Biomass Burning in the UK*. London: Defra, 2017.

12 The wrong transport

1. Curtis, C. (2005), "The windscreen world of land use transport integration: experiences from Perth, WA, a dispersed city." *Town Planning Review*, Vol. 76(4), 423–54.
2. Holman, C., Harrison, R., and Querol, X. (2015), "Review of the efficacy of low emission zones to improve urban air quality in European cities." *Atmospheric Environment*, Vol. 111, 161–9.
3. GEMB mbH and Green-Zones GmbH, "CRIT'Air." https://www.crit-air.fr/en.html.
4. Transport for London, *Travel in London Report 3*. London: TfL, 2010.
5. Ibid.
6. Ellison, R.B., Greaves, S.P., and Hensher, D.A. (2013), "Five years of London's low emission zone: Effects on vehicle fleet composition and air quality." *Transportation Research*, Part D: Transport and Environment, Vol. 23, 25–33.
7. Malina, C., and Scheffler, F. (2015), "The impact of Low Emission

Zones on particulate matter concentration and public health." *Transportation Research*, Part A: Policy and Practice, Vol. 77, 372–85.

8. Boogaard, H., Janssen, N.A., Fischer, P.H., Kos, G.P., Weijers, E.P., Cassee, F.R., van der Zee, S.C., de Hartog, J.J., Meliefste, K., Wang, M., and Brunekreef, B. (2012), "Impact of low emission zones and local traffic policies on ambient air pollution concentrations." *Science of the Total Environment*, Vol. 435, 132–40.
9. Ellison et al., "Five years of London's low emission zone."
10. Wolff, H. (2014), "Keep your clunker in the suburbs: Low emissions zones and the adoption of green vehicles." *The Economic Journal*, Vol. 124, 481–512.
11. Ellison et al., "Five years of London's low emission zone"; Font, A., Guiseppin, L., Ghersi, V., and Fuller, G.W. (2018), "A tale of two cities: Is air pollution improving in London and Paris?" Not yet published.
12. Carslaw, D.C., and Rhys-Tyler, G. (2013), "New insights from comprehensive on-road measurements of NOx, NO2 and NH3 from vehicle emission remote sensing in London, UK." *Atmospheric Environment*, Vol. 81, 339–47.
13. Fuller, G., and Moukhtar, S., "Paris tries something different in the fight against smog." *The Guardian*, January 29, 2017: https://www.theguardian.com/environment/2017/jan/29/paris-fight-against-smog-world-pollutionwatch; Fuller, G., "How different cities responded to December's winter smog." *The Guardian*, January 8, 2017: https://www.theguardian.com/environment/2017/jan/08/how-different-cities-respond-to-winter-smog-pollutionwatch and references therein.
14. Lin, C.Y.C., Zhang, W., and Umanskaya, V.I. (2011), "The effects of driving restrictions on air quality: São Paulo, Bogotá, Beijing, and

Tianjin." Agricultural & Applied Economics Association's 2011 AAEA & NAREA Joint Annual Meeting. Pittsburg, PA; Bigazzi, A.Y., and Rouleau, M. (2017), "Can traffic management strategies improve urban air quality? A review of the evidence." *Journal of Transport & Health*, Vol. 7, 111–24.

15. Kelly, F., Anderson, H.R., Armstrong, B., Atkinson, R., Barratt, B., Beevers, S., Derwent, D., Green, D., Mudway, I., and Wilkinson, P. (2011), "The impact of the congestion charging scheme on air quality in London," Part 1 & 2. Boston, MA: Health Effects Institute.
16. Hanna, R., Kreindler, G., and Olken, B.A. (2017), "Citywide effects of high-occupancy vehicle restrictions: Evidence from 'three-in-one' in Jakarta." *Science*, Vol. 357(6346), 89–93.
17. Jevons, W.S., *The Coal Question: An inquiry concerning the progress of the nation, and the probable exhaustion of our coal mines,* 1st edn. London and Cambridge: Macmillan & Co., 1865.
18. The Standing Advisory Committee on Trunk Road Assessment (Chair: D.A. Woods QC), *Truck Roads and the Generation of Traffic.* London: Department for Transport, 1994.
19. Matson, L., Taylor, T., Sloman, L., and Elliott, J., *Beyond Transport Infrastructure: Lessons for the future from recent road projects*. London: Council for the Protection of Rural England and the Countryside Agency, 2006.
20. Milam, R.T., Birnbaum, M., Ganson, C., Handy, S., and Walters, J. (2017), "Closing the induced vehicle travel gap between research and practice." *Journal of the Transportation Research Board*, Vol. 2653, 10–16.
21. Cairns, S., Atkins, S., and Goodwin, P. (2002), "Disappearing traffic? The story so far." *Proceedings of the Institution of Civil Engineers—Municipal Engineer*, Vol. 151(1), 13–22.

22. Dablanc, L. (2015), "Goods transport in large European cities: Difficult to organize, difficult to modernize." *Transportation Research*, Part A: Policy and Practice, Vol. 41(3), 280–5.
23. Department for Transport, *Road Traffic Estimates: Great Britain 2016*. London: DfT, 2017. https://www.gov.uk/government/uploads/system/uploads/attachment_data/file/611304/annual-road-traffic-estimates-2016.pdf.
24. Transport for London, *Roads Task Force—Technical note 5. What are the main trends and developments affecting van traffic in London?* London: TfL, 2015.
25. Dablanc, L., and Montenon, A. (2015), "Impacts of environmental access restrictions on freight delivery activities: Example of Low Emission Zones in Europe." *Transportation Research Record: Journal of the Transportation Research Board*, Vol. 2478, 12–18.
26. MailRail, "Operation." http://www.mailrail.co.uk/operation.html; Living History of Illinois and Chicago, "Chicago's underground freight railway network." http://livinghistoryofillinois.com/pdf_files/Chicago%20Underground%20Freight%20Railway%20Network.pdf.
27. Asthana, A., and Taylor, M., "Britain to ban sale of all diesel and petrol cars and vans from 2040." *The Guardian*, July 25, 2017: https://www.theguardian.com/politics/2017/jul/25/britain-to-ban-sale-of-all-diesel-and-petrol-cars-and-vans-from-2040.
28. Baynes, C., "Paris to ban all petrol and diesel cars by 2030." *The Guardian*, October 12, 2017: https://www.standard.co.uk/news/world/paris-to-ban-all-combustion-engine-petrol-diesel-cars-by-2030-a3656821.html.
29. Font, A., and Fuller, G.W. (2016), "Did policies to abate atmospheric

emissions from traffic have a positive effect in London?" *Environmental Pollution*, Vol. 218, 463–74.

30. Howard, K., "Disc Brakes." *Motor Sport*, May 2003: https://www.motorsportmagazine.com/archive/article/may-2000/53/disc-brakes.
31. Hagino, H., Oyama, M., and Sasaki, S. (2016), "Laboratory testing of airborne brake wear particle emissions using a dynamometer system under urban city driving cycles." *Atmospheric Environment*, Vol. 131, 269–78.
32. Cassee, F.R., Héroux, M.E., Gerlofs-Nijland, M.E., and Kelly, F.J. (2013), "Particulate matter beyond mass: Recent health evidence on the role of fractions, chemical constituents and sources of emission." *Inhalation Toxicology*, Vol. 25(4), 802–12.
33. Timmers, V.R., and Achten, P.A. (2016), "Non-exhaust PM emissions from electric vehicles." *Atmospheric Environment*, Vol. 134, 10–17.
34. Department for Transport, *Road Traffic Estimates: Great Britain 2016.*
35. Royal College of Physicians and Royal College of Paediatrics and Child Health, *Every Breath We Take: The lifelong impact of air pollution*. London: Royal College of Physicians, 2016.
36. Jarrett, J., Woodcock, J., Griffiths, U.K., Chalabi, Z., Edwards, P., Roberts, I., and Haines, A. (2012), "Effect of increasing active travel in urban England and Wales on costs to the National Health Service." *The Lancet*, Vol. 379(9832), 2198–205.
37. Rojas-Rueda, D., de Nazelle, A., Tainio, M., and Nieuwenhuijsen, M.J. (2011), "The health risks and benefits of cycling in urban environments compared with car use: Health impact assessment study." *British Medical Journal*, Vol. 343, 4521.
38. Rabl, A., and De Nazelle, A. (2011), "Benefits of shift from car to active transport." *Transport Policy*, Vol. 191(1), 121–31.

39. Tainio, M., de Nazelle, A.J., Götschi, T., Kahlmeier, S., Rojas-Rueda, D., Nieuwenhuijsen, M.J., de Sá, T.H., Kelly, P., and Woodcock, J. (2016), "Can air pollution negate the health benefits of cycling and walking?" *Preventive Medicine*, Vol. 87, 233–6.
40. Woodcock, J., Tainio, M., Cheshire, J., O'Brien, O., and Goodman, A. (2013), "Health effects of the London bicycle sharing system: Health impact modelling study." *British Medical Journal,* Vol. 348, 425.
41. UK Biobank, http://www.ukbiobank.ac.uk/.
42. Wheeler, B., "60mph motorway speed limit plan shelved." *BBC News*, July 8, 2014.
43. *The Argus*, "Driving out the motorist: Brighton and Hove anti-car policies slammed by AA." April 12, 2013: http://www.theargus.co.uk/news/10352165. Driving_out_the_motorist_anti_car_policies_slammed_by_AA/.
44. Metz, D. (2013), "Peak car and beyond: The fourth era of travel." *Transport Reviews*, Vol. 33(3), 255–70.
45. Department for Transport, *Road Traffic Estimates: Great Britain 2016.* https://www.gov.uk/government/uploads/system/uploads/attachment_data/file/611304/annual-road-traffic-estimates-2016.pdf.
46. Focas, C., and Christidis, P. (2017), "Peak Car in Europe?" *Transportation Research Procedia*, Vol. 25, 531–50.
47. "Healthy Streets." https://healthystreets.com/.

13 Cleaning the air

1. Evelyn, John, *Fumifugium, or, The inconveniencie of the aer and smoak of London dissipated together with some remedies humbly proposed*, translated by Anna Gross and Justine Shaw. Brighton: Environmental Protection UK, 1661; 2012.

2. Salmond, J.A., Tadaki, M., Vardoulakis, S., Arbuthnott, K., Coutts, A., Demuzere, M., Dirks, K.N., Heaviside, C., Lim, S., Macintyre, H., and McInnes, R.N. (2016), "Health and climate related ecosystem services provided by street trees in the urban environment." *Environmental Health*, Vol. 15, suppl. 1, S36.
3. McDonald, A.G., Bealey, W.J., Fowler, D., Dragosits, U., Skiba, U., Smith, R.I., Donovan, R.G., Brett, H.E., Hewitt, C.N., and Nemitz, E. (2007), "Quantifying the effect of urban tree planting on concentrations and depositions of PM10 in two UK conurbations." *Atmospheric Environment*, Vol. 41(38), 8455–67.
4. Churkina, G., Kuik, F., Bonn, B., Lauer, A., Grote, R., Tomiak, K., and Butler, T.M. (2017), "Effect of VOC emissions from vegetation on air quality in Berlin during a heatwave." *Environmental Science & Technology*, Vol. 51, 6120–30.
5. Lewis, A., "Beware China's 'anti-smog tower' and other plans to pull pollution from the air." *The Conversation*, January 18, 2018.
6. Air Quality Expert Group, *Paints and Surfaces for the Removal of Nitrogen Oxides*. London: Defra, 2016.
7. D'Antoni, D., Smith, L., Auyeung, V., and Weinman, J. (2017), "Psychosocial and demographic predictors of adherence and non-adherence to health advice accompanying air quality warning systems: A systematic review." *Environmental Health*, Vol. 16(1).
8. Lewis, A., and Edwards, P. (2016), "Validate personal air-pollution sensors: Alastair Lewis and Peter Edwards call on researchers to test the accuracy of low-cost monitoring devices before regulators are flooded with questionable air-quality data." *Nature*, Vol. 535(7610), 29–32; Smith, K.R., Edwards, P.M., Evans, M.J., Lee, J.D., Shaw, M.D., Squires, F., Wilde, S., and Lewis, A.C. (2017), "Clustering approaches

to improve the performance of low cost air pollution sensors." *Faraday Discussions*, Vol. 200, 621–37.

9. Laumbach, R., Meng, Q., and Kipen, H. (2015), "What can individuals do to reduce personal health risks from air pollution?" *Journal of Thoracic Disease*, Vol. 7(1), 96.
10. Jones, A.M., Harrison, R.M., Barratt, B., and Fuller, G. (2012), "A large reduction in airborne particle number concentrations at the time of the introduction of 'sulphur free' diesel and the London low emission zone." *Atmospheric Environment*, Vol. 50, 129–38.
11. Kelly, I., and Clancy, L. (1984), "Mortality in a general hospital and urban air pollution." *Irish Medical Journal*, Vol. 77(10), 322–4.
12. Clancy, L., Goodman, P., Sinclair, H., and Dockery, D.W. (2002), "Effect of air-pollution control on death rates in Dublin, Ireland: An intervention study." *The Lancet*, Vol. 360(9341).
13. Dockery, D.W., Rich, D.Q., Goodman, P.G., Clancy, L., Ohman-Strickland, P., George, P., and Kotlov, T. (2013), "Effect of air pollution control on mortality and hospital admissions in Ireland." *Health Effects Institute*, Vol. 176, 3–109.
14. Pozzer, A., Tsimpidi, A.P., Karydis, V.A., De Meij, A., and Lelieveld, J. (2017), "Impact of agricultural emission reductions on fine-particulate matter and public health." *Atmospheric Chemistry and Physics*, 12813.
15. European Union, "Improving air quality: EU acceptance of the Gothenburg Protocol amendment in sight." July 17, 2017: http://www.consilium.europa.eu/en/press/press-releases/2017/07/17/agri-improving-air-quality/.
16. Kumar, A., "Law aiding Monsanto is reason for Delhi's annual smoke season." *The Sunday Guardian Live*, December 30, 2017.

17. Johnston, F.H., Purdie, S., Jalaludin, B., Martin, K.L., Henderson, S.B., and Morgan, G.G. (2014), "Air pollution events from forest fires and emergency department attendances in Sydney, Australia 1996–2007: A case-crossover analysis." *Environmental Heath*, Vol. 13(1), 105.
18. Fuller, G., "Pollutionwatch: sepia skies point to smoke and smog in our atmosphere." *The Guardian*, November 12, 2017: https://www.theguardian.com/uk-news/2017/nov/12/pollutionwatch-sepia-skies-point-to-smoke-and-smog-in-our-atmosphere.
19. Witham, C., and Manning, A. (2007), "Impacts of Russian biomass burning on UK air quality." *Atmospheric Environment*, Vol. 41(37), 8075–90.
20. Johnston, F.H., Henderson, S.B., Chen, Y., Randerson, J.T., Marlier, M., DeFries, R.S., Kinney, P., Bowman, D.M., and Brauer, M. (2012), "Estimated global mortality attributable to smoke from landscape fires." *Environmental Health Perspectives*, Vol. 120(5), 695.

14 A conclusion: What happens next?

1. Stern, N., "The best of centuries or the worst of centuries." Fulbright Commission, June 2018: http://fulbright.org.uk/media/2249/nicholas-stern-essay.pdf.
2. World Bank, *Urban Development—overview*. The World Bank, January 2, 2018: http://www.worldbank.org/en/topic/urbandevelopment/overview.
3. Walton, H., Dajnak, D., Beevers, S., Williams, M., Watkiss, P., and Hunt, A., *Understanding the Health Impacts of Air Pollution in London*. London: King's College London, 2016.
4. Bruckmann, P., Pfeffer, U., and Hoffmann, V. (2014), "50 years of air quality control in Northwestern Germany—how the blue skies over

the Ruhr district were achieved." *Gefahrstoffe-Reinhaltung der Luft*, Vol. 74(1–2), 37–44.

5. Ahlers, A.L. (2015), "How the Sky over the Ruhr Became Blue Again—Or: A German researcher's optimism about China's opportunities to tackle the problem of air pollution." Academia.edu: http://www.academia.edu/17286084/How_the_Sky_over_the_Ruhr_Became_Blue_Again_Or_A_German_researcher_s_optimism_about_China_s_opportunities_to_tackle_the_problem_of_air_pollution_2015_.
6. German Environment Agency, "Federal Environment Agency: The sky over the Ruhr is blue again!" UBA: https://www.umweltbundesamt.de/en/press/pressinformation/federal-environment-agency-sky-over-ruhr-is-blue.
7. Carr, E., and Chan, Y., "Is China serious about curbing pollution along the belt and road?" *China Morning Post*, December 11, 2017.
8. Mayor of London press release, "Sadiq Khan unveils action plan to battle London's toxic air." London.gov, July 5, 2016: https://www.london.gov.uk/press-releases/mayoral/mayor-unveils-action-plan-to-battle-toxic-air.
9. U.S. Environmental Protection Agency Office for Air and Radiation, *The Benefits and Costs of the Clean Air Act from 1990 to 2020*. s.l.: USEPA, 2011.
10. Turnock, S.T., Butt, E.W., Richardson, T.B., Mann, G.W., Reddington, C.L., Forster, P.M., Haywood, J., Crippa, M., Janssens-Maenhout, G., Johnson, C.E., and Bellouin, N. (2016), "The impact of European legislative and technology measures to reduce air pollutants on air." *Environmental Research Letters*, Vol. 12(11), 024010.

11. Department for Transport, *Road Traffic Estimates: Great Britain 2016*. London: DfT, 2017.
12. Fuller, G., "Pollutionwatch: Bicycles take over City of London rush hour." *The Guardian*, April 12, 2018: https://www.theguardian.com/environment/2018/apr/12/pollutionwatch-bicycles-take-over-city-of-london-rush-hour and references therein.
13. Chung, J.H., Hwang, K.Y., and Bae, Y.K. (2012), "The loss of road capacity and self-compliance: Lessons from the Cheonggyecheon stream restoration." *Transport Policy*, Vol. 21, 165–78.
14. McDonald, B.C., de Gouw, J.A., Gilman, J.B., Jathar, S.H., Akherati, A., Cappa, C.D., Jimenez, J.L., Lee-Taylor, J., Hayes, P.L., McKeen, S.A., and Cui, Y.Y. (2018), "Volatile chemical products emerging as largest petrochemical source of urban organic emissions." *Science*, Vol. 359(6377), 760–4.
15. Royal College of Physicians and Royal College of Paediatrics and Child Health, *Every Breath We Take: The lifelong impact of air pollution*. London: Royal College of Physicians, 2016.
16. Hardin, G. (1968), "The Tragedy of the Commons." *Science*, Vol. 162, 1243.
17. Woo, L., "Garrett Hardin, 88; Ecologist Sparked Debate with Controversial Theories." *Los Angeles Times*, September 20, 2003.
18. Li, H., Zhang, Q., Duan, F., Zheng, B., and He, K. (2016), "The 'Parade Blue': Effects of short-term emission control on aerosol chemistry." *Faraday Discussions*, Vol. 189, 317–35.
19. European Commission staff working paper, annex to *The Communication on Thematic Strategy on Air Pollution and the Directive on "Ambient Air Quality and Cleaner Air for Europe."* Brussels: EC, 2005.

20. Amann, M. (ed.), *Final Policy Scenarios of the EU Clean Air Policy Package*. Laxenburg, Austria: International Institute for Applied Systems Analysis, 2014.
21. Song, C., He, J., Wu, L., Jin, T., Chen, X., Li, R., Ren, P., Zhang, L., and Mao, H. (2017), "Health burden attributable to ambient PM2.5 in China." *Environmental Pollution,* Vol. 223, 575–86.
22. The United Nations, Paris Agreement. November 4, 2016: https://unfccc.int/process/conferences/pastconferences/paris-climate-change-conference-november-2015/paris-agreement.
23. Hansen, J., *Storms of My Grandchildren*. London: Bloomsbury, 2009.
24. World Health Organization, *Reducing Global Health Risks through Mitigation of Short-lived Climate Pollutants: Scoping report for policy makers*. Geneva: WHO, 2015.
25. United Nations Environment Programme and World Meteorological Organization, *Integrated Assessment of Black Carbon and Tropospheric Ozone: A summary for policy makers*. Nairobi: UNEP & WMO, 2011; Shindell, D., Kuylenstierna, J.C., Vignati, E., van Dingenen, R., Amann, M., Klimont, Z., Anenberg, S.C., Muller, N., Janssens-Maenhout, G., Raes, F., Schwartz, J., Williams, M., and Fowler, D., (2012), "Simultaneously mitigating near-term climate change and improving human health and food security." *Science,* Vol. 335(6065), 183–9.
26. Williams, M.L., Beevers, S., Kitwiroon, N., Dajnak, D., Walton, H., Lott, M.C., Pye, S., Fecht, D., Toledano, M.B., and Holland, M., *Public Health Air Pollution Impacts of Pathway Options to Meet the 2050 UK Climate Change Act Target—A modelling study*. London: King's College London, 2018.
27. Department for Transport, *The Road to Zero*. London: DfT, 2018.

28. Stern, "The best of centuries or the worst of centuries."
29. Royal College of Physicians and Royal College of Paediatrics and Child Health, *Every Breath We Take: The lifelong impact of air pollution*. London: Royal College of Physicians, 2016.

Index